First contact with TensorFlow
get started with Deep Learning programming

D1420315

First contact with TensorFlow
get started with Deep Learning programming

Jordi Torres

Universitat Politècnica de Catalunya – UPC Barcelona Tech
Barcelona Supercomputing Center – Centro Nacional de Supercomputación (BSC-CNS)

April, 2016

Cover illustration: Supercomputer Marenostrum - Torre Girona chapel

WATCH THIS SPACE Collection

First english edition: March 2016.

☺ Jordi Torres
www.JordiTorres.eu

Universitat Politècncia de Catalunya - UPC Barcelona Tech
UPC Campus Nord, mòdul C6 desp. 217
Jordi Girona 1-3
08034 Barcelona

Cover design: Jordi Torres
Ilustrations: Jordi Torres

Orthographic and typographic proofreader: Laura Juan Merino

Editor: Ferran Julià Massó
Publisher: Jordi Torres , BSC-CNS

Citation:
    First contact with TensorFlow,
    get started with Deep Learing programming
    Jordi Torres,
    Ed. BSC-CNS, Barcelona, 2016
    ISBN 978-1-326-56933-4

This book is devoted to the open-source community,
whose work we consume every day without knowing

Contents

Foreword

The area of Machine Learning has shown a great expansion thanks to the co-development of key areas such as computing, massive data storage and Internet technologies. Many of the technologies and events of everyday life of many people are directly or indirectly influenced by automatic learning. Examples of technologies such as speech recognition, image classification on our phones or detection of spam emails, have enabled apps that a decade ago would have only sounded possible in science fiction. The use of learning in stock market models or medical models has impacted our society massively. In addition, cars with cruise control, drones and robots of all types will impact society in the not too distant future.

Deep Learning, a subtype of Machine Learning, has undoubtedly been one of the fields which has had an explosive expansion since it was rediscovered in 2006. Indeed, many of the startups in Silicon Valley specialize in it, and big technology companies like Google, Facebook, Microsoft or IBM have both development and research teams. Deep Learning has generated interest even outside the university and research areas: a lot of specialized magazines (like *Wired*) and even generic ones (such as *New York Times*, *Bloomberg* or *BBC*) have written many articles about this subject.

This interest has led many students, entrepreneurs and investors to join Deep Learning. Thanks to all the interest generated, several packages have been opened as "Open Source". Being one of the main promoters of the library we developed at Berkeley (Caffe) in 2012 as a PhD student, I can say that TensorFlow, presented in this book and also designed by Google (California), where I have been researching since 2013, will be one of the main tools that researchers and SME companies will use to develop their ideas about Deep Learning and Machine Learning. A guarantee of this is the number of engineers and top researchers who have participated in this project, culminated with the Open Sourcing.

I hope this introductory book will help the reader interested in starting their adventure in this very interesting field. I would like to thank the author, whom I have the pleasure of knowing, the effort to disseminate this technology. He wrote this book (first Spanish version) in record time, two months after the open source project release was announced. This is another example of the vitality of Barcelona and its interest to be one of the actors in this technological scenario that undoubtedly will impact our future.

Oriol Vinyals
Research Scientist at *Google Brain*

Preface

Education is the most powerful weapon which you can use to change the world.

Nelson Mandela

The purpose of this book is to help to spread this knowledge among engineers who want to expand their wisdom in the exciting world of Machine Learning. I believe that anyone with an engineering background may find applications of Deep Learning, and Machine Learning in general, valuable to their work.

Given my background, the reader probably will wonder why I have proposed this challenge of writing about this new Deep Learning technology. My research focus is gradually moving from supercomputing architectures and runtimes to execution middleware's for big data workloads, and more recently to platforms for Machine Learning on massive data.

Precisely by being an engineer, not a data scientist, I think I can contribute with this introductory approach to the subject, and that it can be helpful for many engineers in the early stages; then it will be their choice to go deeper into what they need.

I hope this book adds some value to this world of education that I love so much. I think that knowledge is liberation and should be accessible to all. For this reason, the content of this book will be available on the website *www.JordiTorres.eu/TensorFlow* completely free. If the reader finds the content useful and considers it appropriate to compensate the effort of the author in writing it, there is a tab on the website to make a donation. On the other hand, if the reader prefers to opt for a paper copy, you can purchase the book through *Amazon.com* portal.

A Spanish version is also available. Indeed, this book is the translation of the Spanish one, which was finished last January and it was presented in the GEMLeB Meetup (Grup d'Estudi de Machine Learning de Barcelona) of which I am one of the coorganizers.

Let me thank you for reading this book! It comforts me and justifies my effort for writing it. Those who know me, know that technological diffusion is one of my passions. It energizes and motivates me to keep learning.

Jordi Torres, February 2016

A practical approach

Tell me and I forget. Teach me and I remember. Involve me and I learn.

Benjamin Franklin

One of the common applications of Deep Learning includes pattern recognition. Therefore, in the same way as when you start programming there is sort of a tradition to start printing "Hello World", in Deep Learning a model for the recognition of handwritten digits is usually constructed[1]. The first example of a neural network that I will provide, will also allow me to introduce this new technology called TensorFlow.

However, I do not intend to write a research book on Machine Learning or Deep Learning, I only want to make this new Machine Learning's package, TensorFlow, available to everybody, as soon as possible. Therefore I apologise in to my fellow data scientists for certain simplifications that I have allowed myself in order to share this knowledge with the general reader.

---

[1] The MNIST database of handwritten digits. [Online]. Available at: http://yann.lecun.com/exdb/mnist [Accessed: 16/12/2015].

The reader will find here the regular structure that I use in my classes; that is inviting you to use your computer's keyboard while you learn. We call it *"learn by doing"*, and my experience as a professor at UPC tells me that it is an approach that works very well with engineers who are trying to start a new topic.

For this reason, the book is of a practical nature, and therefore I have reduced the theoretical part as much as possible. However certain mathematical details have been included in the text when they are necessary for the learning process.

I assume that the reader has some basic underestanding of Machine Learning, so I will use some popular algorithms to gradually organize the reader's training in TensorFlow.

In the first chapter, in addition to an introduction to the scenario in which TensorFlow will have an important role, I take the opportunity to explain the basic structure of a TensorFlow program, and explain briefly the data it maintains internally.

In chapter two, through an example of linear regression, I will present some code basics and, at the same time, how to call various important components in the learning process, such as the cost function or the gradient descent optimization algorithm.

In chapter three, where I present a clustering algorithm, I go into detail to present the basic data structure of TensorFlow called *tensor*, and the different classes and functions that the TensorFlow package offers to create and manage the tensors.

In chapter four, how to build a neural network with a single layer to recognize handwritten digits is presented in detail.

This will allow us to sort all the concepts presented above, as well as see the entire process of creating and testing a model.

The next chapter begins with an explanation based on neural network concepts seen in the previous chapter and introduces how to construct a multilayer neural network to get a better result in the recognition of handwritten digits. What it is known as convolutional neural network will be presented in more detail.

In chapter six we look at a more specific issue, probably not of interest to all readers, harnessing the power of calculation presented by GPUs. As introduced in chapter 1, GPUs play an important role in the training process of neural networks.

The book ends with closing remarks, in which I highlight some conclusions. I would like to emphasize that the examples of code in this book can be downloaded from the github repository of the book[2].

---

[2] *Github,* (2016) Fist Contact with TensorFlow. Source code [Online]. Available at: https://github.com/jorditorresBCN/ TutorialTensorFlow [Accessed: 16/12/2015].

1. TensorFlow basics

In this chapter I will present very briefly how a TensorFlow's code and their programming model is. At the end of this chapter, it is expected that the reader can install the Tensor-Flow package on their personal computer.

An Open Source Package

Machine Learning has been investigated by the academy for decades, but it is only in recent years that its penetration has also increased in corporations. This happened thanks to the large volume of data it already had and the unprecedented computing capacity available nowadays.

In this scenario, there is no doubt that Google, under the holding of Alphabet, is one of the largest corporations where Machine Learning technology plays a key role in all of its virtual initiatives and products.

Last October, when Alphabet announced its quarterly Google's results, with considerable increases in sales and profits, CEO Sundar Pichai said clearly: *"Machine learning is a core, transformative way by which we're rethinking everything we're doing"*.

Technologically speaking, we are facing a change of era in which Google is not the only big player. Other technology companies such as Microsoft, Facebook, Amazon and Apple, among many other corporations are also increasing their investment in these areas.

In this context, a few months ago Google released its Tensor-Flow engine under an open source license (Apache 2.0). TensorFlow can be used by developers and researchers who want to incorporate Machine Learning in their projects and products, in the same way that Google is doing internally with different commercial products like Gmail, Google Photos, Search, voice recognition, etc.

TensorFlow was originally developed by the Google Brain Team, with the purpose of conducting Machine Learning and deep neural networks research, but the system is general enough to be applied in a wide variety of other Machine Learning problems.

Since I am an engineer and I am speaking to engineers, the book will look under the hood to see how the algorithms are represented by a data flow graph. TensorFlow can be seen as a library for numerical computation using data flow graphs. The nodes in the graph represent mathematical operations, while the graph edges represent the multidimensional data arrays (tensors), which interconnect the nodes.

TensorFlow is constructed around the basic idea of building and manipulating a computational graph, representing symbolically the numerical operations to be performed. This allows TensorFlow to take advantage of both CPUs and GPUs right now from Linux 64-bit platforms such as Mac OS X, as well as mobile platforms such as Android or iOS.

Another strength of this new package is its visual Tensor-Board module that allows a lot of information about how the algorithm is running to be monitored and displayed. Being able to measure and display the behavior of algorithms is extremely important in the process of creating better models. I have a feeling that currently many models are refined through a little blind process, through trial and error, with the obvious waste of resources and, above all, time.

TensorFlow Serving

Recently Google launched TensorFlow Serving[3], that helps developers to take their TensorFlow machine learning models (and, even so, can be extended to serve other types of models) into production. TensorFlow Serving is an open source serving system (written in C++) now available on GitHub under the Apache 2.0 license.

What is the difference between TensorFlow and TensorFlow Serving? While in TensorFlow it is easier for the developers to build machine learning algorithms and train them for certain types of data inputs, TensorFlow Serving specializes in making these models usable in production environments. The idea is that developers train their models using TensorFlow and then they use TensorFlow Serving's APIs to react to input from a client.

---

[3] *TensorFlow Serving* [Online]. Available at:
http://tensorflow.github.io/serving/ [Accessed: 24/02/2016].

This allows developers to experiment with different models on a large scale that change over time, based on real-world data, and maintain a stable architecture and API in place.

The typical pipeline is that a training data is fed to the learner, which outputs a model, which after being validated is ready to be deployed to the TensorFlow serving system. It is quite common to launch and iterate on our model over time, as new data becomes available, or as you improve the model. In fact, in the google post[4] they mention that at Google, many pipelines are running continuously, producing new model versions as new data becomes available.

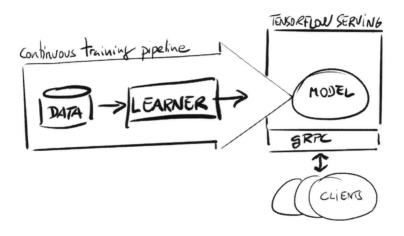

4   Google Research Blog [Online]. Available at:
http://googleresearch.blogspot.com.es/2016/02/running-your-models-in-production-with.html?m=1[Accessed: 24/02/2016].

Developers use to communicate with TensorFlow Serving a front-end implementation based on gRPC, a high performance, open source RPC framework from Google.

If you are interested in learning more about TensorFlow Serving, I suggest you start by by reading the Serving architecture overview[5] section, set up your environment and start to do a basic tutorial[6] .

TensorFlow Installation

It is time to get your hands dirty. From now on, I recommend that you interleave the reading with the practice on your computer.

TensorFlow has a Python API (plus a C / C ++) that requires the installation of Python 2.7 (I assume that any engineer who reads this book knows how to do it).

In general, when you are working in Python, you should use the virtual environment *virtualenv*. *Virtualenv* is a tool to keep Python dependencies required in different projects, in different parts of the same computer. If we use *virtualenv* to install TensorFlow, this will not overwrite existing versions of Python packages from other projects required by TensorFlow.

---

[5] *TensorFlow Serving*-Architecture Overview[Online]. Available at: http://tensorflow.github.io/serving/[Accessed: 24/02/2016].
[6]*TensorFlow Serving*- Serving a TensorFlow Model [Online]. Available at: http://tensorflow.github.io/serving/serving_basic [Accessed: 24/02/2016].

First, you should install *pip* and *virtualenv* if they are not already installed, like the follow script shows:

```
# Ubuntu/Linux 64-bit
$ sudo apt-get install python-pip python-dev python-virtualenv

# Mac OS X
$ sudo easy_install pip
$ sudo pip install --upgrade virtualenv
```

Then you must create a virtual environment *virtualenv*. The following commands create a virtual environment *virtualenv* in the ~/tensorflow directory:

```
$ virtualenv --system-site-packages ~/tensorflow
```

The next step is to activate the virtualenv. This can be done as follows:

```
$ source ~/tensorflow/bin/activate# si se usa bash
$ source ~/tensorflow/bin/activate.csh# si se usa csh
(tensorflow)$
```

The name of the virtual environment in which we are working will appear at the beginning of each command line from now on. Once the *virtualenv* is activated, you can use *pip* to install TensorFlow inside it:

```
# Ubuntu/Linux 64-bit, CPU only:
(tensorflow)$ sudo pip install --upgrade
https://storage.googleapis.com/tensorflow/linux/cpu/tensorflow
-0.7.1-cp27-none-linux_x86_64.whl
```

```
# Mac OS X, CPU only:
(tensorflow)$  sudo easy_install --upgrade six
(tensorflow)$ sudo pip install --upgrade
https://storage.googleapis.com/tensorflow/mac/tensorflow-
0.7.1-cp27-none-any.whl
```

I recommend that you visit the official documentation indicated here, to be sure that you are installing the latest available version.

If the platform where you are running your code has a GPU, the package to use will be different. I recommend that you visit the official documentation to see if your GPU meets the specifications required to support Tensorflow. Installing additional software is required to run Tensorflow GPU and all the information can be found at *Download and Setup Tensorflow*[7] web page. For more information on the use of GPUs, I suggest reading chapter 6.

Finally, when you've finished, you should disable the virtual environment as follows:

```
(tensorflow)$ deactivate
```

Given the introductory nature of this book, we suggest that the reader visits the mentioned official documentation page to find more information about other ways to install Tensorflow.

---

[7] TensorFlow, (2016) Download & Setup [Online]. Available at: https://www.tensorflow.org/versions/master/get_started/os_setup .html#download-and-setup [Accessed: 16/12/2015].

My first code in TensorFlow

As I mentioned at the beginning, we will move in this exploration of the planet TensorFlow with little theory and lots of practice. Let's start!

From now on, it is best to use any text editor to write python code and save it with extension ".py" (eg *test.py*). To run the code, it will be enough with the command *python test.py*.

To get a first impression of what a TensorFlow's program is, I suggest doing a simple multiplication program; the code looks like this:

```
import tensorflow as tf

a = tf.placeholder("float")
b = tf.placeholder("float")

y = tf.mul(a, b)

sess = tf.Session()

printsess.run(y, feed_dict={a: 3, b: 3})
```

In this code, after importing the Python module tensorflow, we define "symbolic" variables, called *placeholder* in order to manipulate them during the program execution. Then, we move these variables as a parameter in the call to the function multiply that TensorFlow offers. *tf.mul* is one of the many mathematical operations that TensorFlow offers to manipulate the *tensors*. In this moment, tensors can be considered dynamically-sized, multidimensional data arrays.

The main ones are shown in the following table:

Operation	Description
tf.add	sum
tf.sub	substraction
tf.mul	multiplication
tf.div	division
tf.mod	module
tf.abs	return the absolute value
tf.neg	return negative value
tf.sign	return the sign
tf.inv	returns the inverse
tf.square	calculates the square
tf.round	returns the nearest integer
tf.sqrt	calculates the square root
tf.pow	calculates the power
tf.exp	calculates the exponential
tf.log	calculates the logarithm
tf.maximum	returns the maximum
tf.minimum	returns the minimum
tf.cos	calculates the cosine
tf.sin	calculates the sine

TensorFlow also offers the programmer a number of functions to perform mathematical operations on matrices. Some are listed below:

Operation	Description
tf.diag	returns a diagonal tensor with a given diagonal values
tf.transpose	returns the transposes of the argument
tf.matmul	returns a tensor product of multiplying two tensors listed as arguments
tf.matrix_determinant	returns the determinant of the square matrix specified as an argument
tf.matrix_inverse	returns the inverse of the square matrix specified as an argument

The next step, one of the most important, is to create a session to evaluate the specified symbolic expression. Indeed, until now nothing has yet been executed in this TensorFlow-code. Let me emphasize that TensorFlow is both, an interface to express Machine Learning's algorithms and an implementation to run them, and this is a good example.

Programs interact with Tensorflow libraries by creating a session with *Session()*; it is only from the creation of this session when we can call the *run()* method, and that is when it really starts to run the specified code. In this particular example, the values of the variables are introduced into the *run()* method with a *feed_dict* argument. That's when the associated code solves the expression and exits from the display a 9 as a result of multiplication.

With this simple example, I tried to introduce the idea that the normal way to program in TensorFlow is to specify the whole problem first, and eventually create a session to allow the running of the associated computation.

Sometimes however, we are interested in having more flexibility in order to structure the code, inserting operations to build the graph with operations running part of it. It happens when we are, for example, using interactive environments of Python such as IPython[8]. For this purpose, TesorFlow offers the *tf.InteractiveSession()* class.

The motivation for this programming model is beyond the reach of this book. However, to continue with the next chapter, we only need to know that all information is saved internally in a graph structure that contains all the information operations and data .

---

[8] Wikipedia, (2016). IPython. [Online]. Available at:
https://en.wikipedia.org/wiki/IPython
[Accessed: 19/03/2016].

This graph describes mathematical computations. The nodes typically implement mathematical operations, but they can also represent points of data entry, output results, or read/write persistent variables. The edges describe the relationships between nodes with their inputs and outputs and at the same time carry tensors, the basic data structure of TensorFlow.

The representation of the information as a graph allows TensorFlow to know the dependencies between transactions and assigns operations to devices asynchronously, and in parallel, when these operations already have their associated tensors (indicated in the edges input) available.

Parallelism is therefore one of the factors that enables us to speed up the execution of some computationally expensive algorithms, but also because TensorFlow has already efficiently implemented a set of complex operations. In addition, most of these operations have associated kernels which are implementations of operations designed for specific devices such as GPUs. The following table summarizes the most important operations/kernels[9]:

---

[9]   TensorFlow: Large-scale machine learning on heterogeneous systems, (2015). [Online]. Available at: http://download.tensorflow.org/paper/ whitepaper2015.pdf [Accessed: 20/12/2015].

Operations groups	Operations
Maths	Add, Sub, Mul, Div, Exp, Log, Greater, Less, Equal
Array	Concat, Slice, Split, Constant, Rank, Shape, Shuffle
Matrix	MatMul, MatrixInverse, MatrixDeterminant
Neuronal Network	SoftMax, Sigmoid, ReLU, Convolution2D, MaxPool
Checkpointing	Save, Restore
Queues and syncronizations	Enqueue, Dequeue, MutexAcquire, MutexRelease
Flow control	Merge, Switch, Enter, Leave, NextIteration

Display panel Tensorboard

To make it more comprehensive, TensorFlow includes functions to debug and optimize programs in a visualization tool called TensorBoard. TensorBoard can view different types of statistics about the parameters and details of any part of the graph computing graphically.

The data displayed with TensorBoard module is generated during the execution of TensorFlow and stored in trace files whose data is obtained from the *summary operations*. In the documentation page[10] of TensorFlow, you can find detailed explanation of the Python API.

The way we can invoke it is very simple: a service with Tensorflow commands from the command line, which will include as an argument the file that contains the trace.

```
(tensorflow)$ tensorboard --logdir=<trace file>
```

You simply need to access the local socket 6006 from the browser[11] with http://localhost:6006/ .

---

[10] TensorFlow, (2016) *Python API – Summary Operations*. [Online]. Available at: https://www.tensorflow.org/versions/master/api_docs/python/train.html#summary-operations [Accessed: 03/01/2016].

[11] I recommend using Google Chrome to ensure proper display.

The visualization tool called TensorBoard is beyond the reach of this book. For more details about how Tensorboard works, the reader can visit the section *TensorBoard Graph Visualization*[12]from the TensorFlow tutorial page.

---

[12] TensorFlow, (2016) TensorBoard: Graph Visualization.[Online]. Available at: https://www.tensorflow.org/versions/master/ how_tos/graph_viz/index.html[Accessed: 02/01/2016].

2. Linear Regression in TensorFlow

In this chapter, I will begin exploring TensorFlow's coding with a simple model: Linear Regression. Based on this example, I will present some code basics and, at the same time, how to call various important components in the learning process, such as the cost function or the algorithm gradient descent.

Model of relationship between variables

Linear regression is a statistical technique used to measure the relationship between variables. Its interest is that the algorithm that implements it is not conceptually complex, and can also be adapted to a wide variety of situations. For these reasons, I have found it interesting to start delving into TensorFlow with an example of linear regression.

Remember that both, in the case of two variables (simple regression) and the case of more than two variables (multiple regression), linear regression models the relationship between a dependent variable, independent variables x_i and a random term b.

In this section I will create a simple example to explain how TensorFlow works assuming that our data model corresponds to a simple linear regression as $y = W * x + b$. For this,

I use a simple Python program that creates data in a two-dimensional space, and then I will ask TensorFlow to look for the line that fits the best in these points.

The first thing to do is to import the NumPy package that we will use to generate points. The code we have created is as it follows:

```
import numpy as np

num_points = 1000
vectors_set = []
for i in xrange(num_points):
        x1= np.random.normal(0.0, 0.55)
        y1= x1 * 0.1 + 0.3 + np.random.normal(0.0, 0.03)
        vectors_set.append([x1, y1])

x_data = [v[0] for v in vectors_set]
y_data = [v[1] for v in vectors_set]
```

As you can see from the code, we have generated points following the relationship $y = 0.1 * x + 0.3$, albeit with some variation, using a normal distribution, so the points do not fully correspond to a line, allowing us to make a more interesting example.

In our case, a display of the resulting cloud of points is:

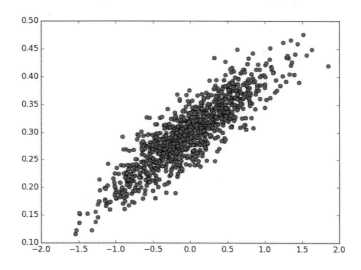

The reader can view them with the following code (in this case, we need to import some of the functions of *matplotlib* package, running *pip install matplotlib*[13]):

```
import matplotlib.pyplot as plt

plt.plot(x_data, y_data, 'ro', label='Original data')
plt.legend()
plt.show()
```

---

[13] One reviewer of this book has indicated that he also had to install the package python-gi-cairo.

These points are the data that we will consider the training dataset for our model.

Cost function and gradient descent algorithm

The next step is to train our learning algorithm to be able to obtain output values y, estimated from the input data x_data. In this case, as we know in advance that it is a linear regression, we can represent our model with only two parameters: W and b.

The objective is to generate a TensorFlow code that allows to find the best parameters W and b, that from input data x_data, adjunct them to y_data output data, in our case it will be a straight line defined by $y\_data = W * x\_data + b$. The reader knows that W should be close to 0.1 and b to 0.3, but TensorFlow does not know and it must realize it for itself.

A standard way to solve such problems is to iterate through each value of the data set and modify the parameters W and b in order to get a more precise answer every time. To find out if we are improving in these iterations, we will define a cost function (also called "error function") that measures how "good" (actually, as "bad") a certain line is.

This function receives the pair of W and as parameters b and returns an error value based on how well the line fits the data. In our example we can use as a cost function the *mean squared error*[14]. With the *mean squared error* we get the average of the "errors" based on the distance between the real values and the estimated one on each iteration of the algorithm.

Later, I will go into more detail with the cost function and its alternatives, but for this introductory example the *mean squared error* helps us to move forward step by step.

Now it is time to program everything that I have explained with TensorFlow. To do this, first we will create three variables with the following sentences:

```
W = tf.Variable(tf.random_uniform([1], -1.0, 1.0))
b = tf.Variable(tf.zeros([1]))
y = W * x_data + b
```

For now, we can move forward knowing only that the call to the method *Variable* is defining a variable that resides in the internal graph data structure of TensorFlow, of which I have spoken above. We will return with more information about the method parameters later, but for now I think that it's better to move forward to facilitate this first approach.

Now, with these variables defined, we can express the cost function that we discussed earlier, based on the distance between each point and the calculated point with the function y= $W * x + b$. After that, we can calculate its square, and aver-

---

[14]   Wikipedia, (2016). Mean Square Error. [Online]. Available at: https://en.wikipedia.org/wiki/Mean_squared_error [Accessed: 9/01/2016].

age the sum. In TensorFlow this cost function is expressed as follows:

```
loss = tf.reduce_mean(tf.square(y - y_data))
```

As we see, this expression calculates the average of the squared distances between the *y_data* point that we know, and the point *y* calculated from the input *x_data*.

At this point, the reader might already suspects that the line that best fits our data is the one that obtains the lesser error value. Therefore, if we minimize the error function, we will find the best model for our data.

Without going into too much detail at the moment, this is what the optimization algorithm that minimizes functions known as *gradient descent*[15] achieves. At a theoretical level gradient descent is an algorithm that given a function defined by a set of parameters, it starts with an initial set of parameter values and iteratively moves toward a set of values that minimize the function. This iterative minimization is achieved taking steps in the negative direction of the function *gradient*[16]. It's conventional to square the distance to en-

---

[15]   Wikipedia, (2016). Gradient descent. [Online]. Available at:
https://en.wikipedia.org/wiki/Gradient_descent
[Accessed: 9/01/2016].
[16]   Wikipedia, (2016). Gradient. [Online]. Available at:
https://en.wikipedia.org/wiki/Gradient [Accessed: 9/01/2016].

sure that it is positive and to make the error function differentiable in order to compute the gradient.

The algorithm begins with the initial values of a set of parameters (in our case W and b), and then the algorithm is iteratively adjusting the value of those variables in a way that, in the end of the process, the values of the variables minimize the cost function.

To use this algorithm in TensorFlow, we just have to execute the following two statements:

```
optimizer = tf.train.GradientDescentOptimizer(0.5)
train = optimizer.minimize(loss)
```

Right now, this is enough to have the idea that TensorFlow has created the relevant data in its internal data structure, and it has also implemented in this structure an optimizer that may be invoked by *train*, which it is a *gradient descent* algorithm to the cost function defined. Later on, we will discuss the function parameter called *learning rate* (in our example with value 0.5).

Running the algorithm

As we have seen before, at this point in the code the calls specified to the library TensorFlow have only added information to its internal graph, and the runtime of TensorFlow has not yet run any of the algorithms. Therefore, like the example of the previous chapter, we must create a *session*, call the *run* method and passing *train* as parameter. Also, because

in the code we have specified variables, we must initialize
them previously with the following calls:

```
init = tf.initialize_all_variables()

sess = tf.Session()
sess.run(init)
```

Now we can start the iterative process that will allow us to
find the values of W and b, defining the model line that best
fits the points of entry. The training process continues until
the model achieves a desired level of accuracy on the training data. In our particular example, if we assume that with
only 8 iterations is sufficient, the code could be:

```
for step in xrange(8):
    sess.run(train)
print step, sess.run(W), sess.run(b)
```

The result of running this code show that the values of W
and b are close to the value that we know beforehand. In my
case, the result of the *print* is:

```
(array([ 0.09150752], dtype=float32), array([ 0.30007562],
dtype=float32))
```

And, if we graphically display the result with the following
code:

```
plt.plot(x_data, y_data, 'ro')
plt.plot(x_data, sess.run(W) * x_data + sess.run(b))
plt.legend()
plt.show()
```

We can see graphically the line defined by parameters $W = 0.0854$ and $b = 0.299$ achieved with only 8 iterations:

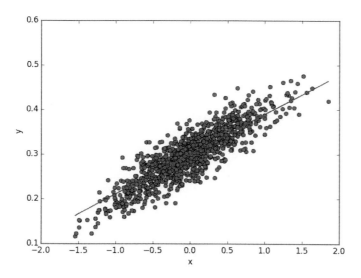

Note that we have only executed eight iterations to simplify the explanation, but if we run more, the value of parameters get closer to the expected values. We can use the following sentence to print the values of W and b:

```
print(step, sess.run(W), sess.run(b))
```

In our case the print outputs are:

```
(0, array([-0.04841119], dtype=float32), array([ 0.29720169], dtype=float32))
(1, array([-0.00449257], dtype=float32), array([ 0.29804006], dtype=float32))
(2, array([ 0.02618564], dtype=float32), array([ 0.29869056], dtype=float32))
(3, array([ 0.04761609], dtype=float32), array([ 0.29914495], dtype=float32))
(4, array([ 0.06258646], dtype=float32), array([ 0.29946238], dtype=float32))
(5, array([ 0.07304412], dtype=float32), array([ 0.29968411], dtype=float32))
(6, array([ 0.08034936], dtype=float32), array([ 0.29983902], dtype=float32))
(7, array([ 0.08545248], dtype=float32), array([ 0.29994723], dtype=float32))
```

You can observe that the algorithm begins with the initial values of $W = -0.0484$ and $b = 0.2972$ (in our case) and then the algorithm is iteratively adjusting in a way that the values of the variables minimize the cost function.

You can also check that the cost function is decreasing with

```
print(step, sess.run(loss))
```

In this case the print output is:

```
(0, 0.015878126)
(1, 0.0079048825)
(2, 0.0041520335)
(3, 0.0023856456)
(4, 0.0015542418)
(5, 0.001162916)
(6, 0.00097872759)
(7, 0.00089203351)
```

I suggest that reader visualizes the plot at each iteration, allowing us to visually observe how the algorithm is adjusting the parameter values. In our case the 8 snapshots are:

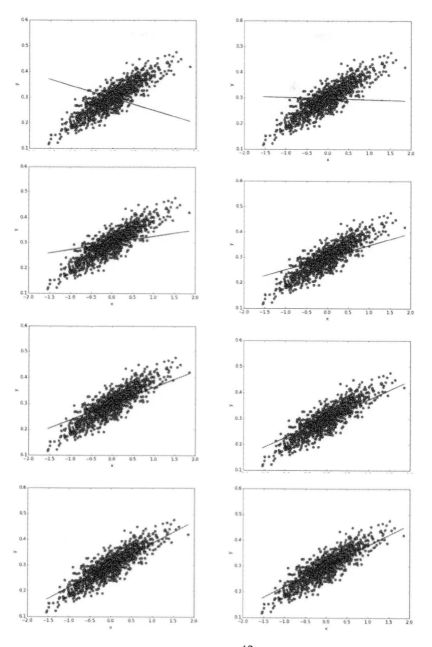

As the reader can see, at each iteration of the algorithm the line fits better to the data. How does the gradient descent algorithm get closer to the values of the parameters that minimize the cost function?

Since our error function consists of two parameters (W and b) we can visualize it as a two-dimensional surface. Each point in this two-dimensional space represents a line. The height of the function at each point is the error value for that line. In this surface some lines yield smaller error values than others. When TensorFlow runs gradient descent search, it will start from some location on this surface (in our example the point $W= -0.04841119$ and $b=0.29720169$) and move downhill to find the line with the lowest error.

To run gradient descent on this error function, TensorFlow computes its gradient. The gradient will act like a compass and always point us downhill. To compute it, TensorFlow will differentiate the error function, that in our case means that it will need to compute a partial derivative for W and b that indicates the direction to move in for each iteration.

The *learning rate* parameter mentioned before, controls how large of a step TensorFlow will take downhill during each iteration. If we introduce a parameter too large of a step, we may step over the minimum. However, if we indicate to TensorFlow to take small steps, it will require much iteration to arrive at the minimum. So using a good learning rate is crucial. There are different techniques to adapt the value of the learning rate parameter, however it is beyond the scope of this introductory book. A good way to ensure that gradient descent algorithm is working fine is to make sure that the error decreases at each iteration.

Remember that in order to facilitate the reader to test the code described in this chapter, you can download it from *Github*[17] of the book with the name of *regression.py*. Here you will find all together for easy tracking:

```
import numpy as np


num_points = 1000
vectors_set = []
for i in xrange(num_points):
        x1= np.random.normal(0.0, 0.55)
        y1= x1 * 0.1 + 0.3 + np.random.normal(0.0, 0.03)
        vectors_set.append([x1, y1])

x_data = [v[0] for v in vectors_set]
y_data = [v[1] for v in vectors_set]



import matplotlib.pyplot as plt

#Graphic display
plt.plot(x_data, y_data, 'ro')
plt.legend()
plt.show()


import tensorflow as tf


W = tf.Variable(tf.random_uniform([1], -1.0, 1.0))
b = tf.Variable(tf.zeros([1]))
y = W * x_data + b
```

---

[17]   *Github*, (2016) Book source code [Online]. Available at: https://github.com/jorditorresBCN/ TutorialTensorFlow . [Accessed: 16/12/2015].

```
loss = tf.reduce_mean(tf.square(y - y_data))
optimizer = tf.train.GradientDescentOptimizer(0.5)
train = optimizer.minimize(loss)

init = tf.initialize_all_variables()

sess = tf.Session()
sess.run(init)

for step in xrange(8):
        sess.run(train)
        print(step, sess.run(W), sess.run(b))
        print(step, sess.run(loss))
        #Graphic display
        plt.plot(x_data, y_data, 'ro')
        plt.plot(x_data, sess.run(W) * x_data + sess.run(b))
        plt.xlabel('x')
        plt.xlim(-2,2)
        plt.ylim(0.1,0.6)
        plt.ylabel('y')
        plt.legend()
        plt.show()
```

In this chapter we have begun to explore the possibilities of the TensorFlow package with a first intuitive approach to two fundamental pieces: the *cost function* and *gradient descent algorithm*, using a basic linear regression algorithm for their introduction. In the next chapter we will go into more detail about the data structures used by TensorFlow package.

3. Clustering in TensorFlow

Linear regression, which has been presented in the previous chapter, is a supervised learning algorithm in which we use the data and output values (or labels) to build a model that fits them. But we haven't always tagged data, and despite this we also want analyze them in some way. In this case, we can use an unsupervised learning algorithm as clustering. The clustering method is widely used because it is often a good approach for preliminary screening data analysis.

In this chapter, I will present the clustering algorithm called *K-means*. It is surely the most popular and widely used to automatically group the data into coherent subsets so that all the elements in a subset are more similar to each other than with the rest. In this algorithm, we do not have any target or outcome variable to predict estimations.

I will also use this chapter to achieve progress in the knowledge of TensorFlow and go into more detail in the basic data structure called *tensor*. I will start by explaining what this type of data is like and present the transformations that can be performed on it. Then, I will show the use of *K-means* algorithm in a case study using *tensors*.

Basic data structure: *tensor*

TensorFlow programs use a basic data structure called *tensor* to represent all of their datum. A tensor can be considered a dynamically-sized multidimensional data arrays that have as a properties a static data type, which can be from *boolean* or *string* to a variety of numeric types. Below is a table of the main types and their equivalent in Python.

Type in TensorFlow	Type in Python	Description
DT_FLOAT	tf.float32	Floating point of 32 bits
DT_INT16	tf.int16	Integer of 16 bits
DT_INT32	tf.int32	Integer of 32 bits
DT_INT64	tf.int64	Integer of 64 bits
DT_STRING	tf.string	String
DT_BOOL	tf.bool	Boolean

In addition, each *tensor* has a *rank*, which is the number of its dimensions. For example, the following tensor (defined as a list in Python) has rank 2:

t = [[1, 2, 3], [4, 5, 6], [7, 8, 9]]

Tensors can have any rank. A rank 2 tensor is usually considered a matrix, and a rank 1 tensor would be a vector. Rank 0 is considered a scalar value.

TensorFlow documentation uses three types of naming conventions to describe the dimension of a tensor: *Shape*, *Rank* and *Dimension Number*. The following table shows the relationship between them in order to make easier the Tensor Flow documentation's traking easier:

Shape	Rank	Dimension Number
[]	0	0-D
[D0]	1	1-D
[D0, D1]	2	2-D
[D0, D1, D2]	3	3-D
...	...	...
[D0, D1, ... Dn]	n	n-D

These tensors can be manipulated with a series of transformations that supply the TensorFlow package. Below, we discuss some of them in the next table.

Throughout this chapter we will go into more detail on some of them. A comprehensive list of transformations and details of each one can be found on the official website of Tensor-Flow, *Tensor Transformations*[18].

---

[18]   TensorFlow, (2016) API de Python - Tensor Transformations [Online]. Available at: https://www.tensorflow.org /versions/master/api_docs/python/ array_ops.html [Accessed: 16/12/2015].

Operation	Description
tf.shape	To find a shape of a *tensor*
tf.size	To find the size of a *tensor*
tf.rank	To find a rank of a *tensor*
tf.reshape	To change the shape of a *tensor* keeping the same elements contained
tf.squeeze	To delete in a *tensor* dimensions of size 1
tf.expand_dims	To insert a dimension to a *tensor*
tf.slice	To remove a portions of a *tensor*
tf.split	To divide a *tensor* into several tensors along one dimension
tf.tile	To create a new *tensor* replicating a *tensor* multiple times
tf.concat	To concatenate *tensors* in one dimension
tf.reverse	To reverse a specific dimension of a *tensor*
tf.transpose	To transpose dimensions in a *tensor*
tf.gather	To collect portions according to an index

For example, suppose that you want to extend an array of 2x2000 (a 2D tensor) to a cube (3D tensor). We can use the *tf.expand_ dims* function, which allows us to insert a dimension to a tensor:

```
vectors = tf.constant(vectors_set)
extended_vectors = tf.expand_dims(vectors, 0)
```

In this case, *tf.expand_dims* inserts a dimension into a tensor in the one given in the argument (the dimensions start at zero).

Visually, the above transformation is as follows:

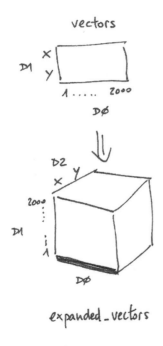

expanded_vectors

As you can see, we now have a 3D *tensor*, but we cannot determine the size of the new dimension D0 based on function arguments.

If we obtain the shape of this *tensor* with the *get_shape()* operation, we can see that there is no associated size:

```
print expanded_vectors.get_shape()
```

It appears on the screen like:

```
TensorShape([Dimension(1), Dimension(2000), Dimension(2)])
```

Later in this chapter, we will see that, thanks to TensorFlow *shape broadcasting,* many mathematical manipulation functions of *tensors* (as presented in the first chapter), are able to discover for themselves the size in the dimension which unspecific size and assign to it this deduced value.

Data Storage in TensorFlow

Following the presentation of TensorFlow's package, broadly speaking there are three main ways of obtaining data on a TensorFlow program:

1. From data files.

2. Data preloaded as constants or variables.

3. Those provided by Python code.

Below, I briefly describe each of them.

1. Data files

Usually, the initial data is downloaded from a data file. The process is not complex, and given the introductory nature of this book I invite the reader to visit the website of Tensor-Flow[19] for more details on how to download data from different file types. You can also review the Python code *input_data.py*[20](available on the Github book), which loads the MNIST data from files (I will use this in the following chapters).

2. Variables and constants

When it comes to small sets, data can also be found pre-loaded into memory; there are two basic ways to create them, as we have seen in the previous example:

- As a constants using *tf.constant(...)*

- As a variable using *tf.Variable(...)*

---

[19]   TensorFlow, (2016) Tutorial – Reading Data [Online]. Available at:https://www.tensorflow.org/versions/master/how_tos/reading_data [Accessed: 16/12/2015].

[20]   *Github*, (2016) TensorFlow Book – Jordi Torres. [Online]. Available at: https://github.com/jorditorresBCN/Libro TensorFlow/blob/master/ input_data.py [Accessed: 19/02/2016].

TensorFlow package offers different operations that can be used to generate constants. In the table below you can find a summary of the most important:

Operation	Description
tf.zeros_like	Creates a tensor with all elements initialized to 0
tf.ones_like	Creates a tensor with all elements initialized to 1
tf.fill	Creates a tensor with all elements initialized to a scalar value given as argument
tf.constant	Creates a tensor of constants with the elements listed as an arguments

In TensorFlow, during the training process of the models, the parameters are maintained in the memory as variables. When a variable is created, you can use a tensor defined as a parameter of the function as an initial value, which can be a constant or a random value. TensorFlow offers a collection of operations that produce random tensors with different distributions:

Operation	Description
tf.random_normal	Random values with a normal distribution
tf.truncated_normal	Random values with a normal distribution but eliminating those values whose magnitude is more than 2 times the standard deviation
tf.random_uniform	Random values with a uniform distribution
tf.random_shuffle	Randomly mixed tensor elements in the first dimension
tf.set_random_seed	Sets the random seed

An important detail is that all of these operations require a specific shape of the tensors as the parameters of the function, and the variable that is created has the same shape. In general, the variables have a fixed shape, but TensorFlow provides mechanisms to reshape it if necessary.

When using variables, these must be explicitly initialized after the graph that has been constructed, and before any operation is executed with the *run()* function. As we have seen, it can be used *tf.initialize_all_variables()* for this purpose. Variables also can be saved onto disk during and after training model through TensorFlow *tf.train.Saver()* class, but this class is beyond the scope of this book.

3. Provided by Python code

Finally, we can use what we have called "symbolic variable" or *placeholder* to manipulate data during program execution. The call is *placeholder()*, which includes arguments with the type of the elements and the shape of the tensor, and optionally a name.

At the same time as making the calls to *Session.run()* or *Tensor.eval()* from the Python code, this tensor is populated with the data specified in the *feed_dict* parameter. Remember the first code in Chapter 1:

```
import tensorflow as tf
a = tf.placeholder("float")
b = tf.placeholder("float")
y = tf.mul(a, b)
sess = tf.Session()
print  sess.run(y, feed_dict={a: 3, b: 3})
```

In the last line of code, when the call *sess.run()* is made, it is when we pass the values of the two tensors *a* and *b* through *feed_dict* parameter.

With this brief introduction about tensors, I hope that from now on the reader can follow the codes of the following chapters without any difficulty.

K-means algorithm

K-means is a type of unsupervised algorithm which solves the clustering problem. Its procedure follows a simple and easy way to classify a given data set through a certain number of clusters (assume k clusters). Data points inside a cluster are homogeneous and heterogeneous to peer groups, that means that all the elements in a subset are more similar to each other than with the rest.

The result of the algorithm is a set of K dots, called centroids, which are the focus of the different groups obtained, and the tag that represents the set of points that are assigned to only one of the K clusters. All the points within a cluster are closer in distance to the centroid than any of the other centroids.

Making clusters is a computationally expensive problem if we want to minimize the error function directly (what is known as an *NP-hard* problem); and therefore it some algorithms that converge rapidly in a local optimum by heuristics have been created. The most commonly used algorithm uses an iterative refinement technique, which converges in a few iterations.

Broadly speaking, this technique has three steps:

- **Initial step** (step 0): determines an initial set of K centroids.
- **Allocation step** (step 1): assigns each observation to the nearest group.
- **Update step** (step 2): calculates the new centroids for each new group.

There are several methods to determine initial K centroids. One of them is randomly choose K observations in the data set and consider them centroids; this is the one we will use in our example.

The steps of allocation (step 1) and updating (step 2) are being alternated in a loop until it is considered that the algorithm has converged, which may be for example when allocations of points to groups no longer change.

Since this is a heuristic algorithm, there is no guarantee that it converges to the global optimum, and the outcome depends on the initial groups. Therefore, as the algorithm is generally very fast, it is usual to repeat executions multiple times with different values of the initials centroides, and then weigh the result.

To start coding our example of *K-means* in TensorFlow I suggest to first generate some data as a testbed. I propose to do something simple, like generating 2,000 points in a 2D space in a random manner, following two normal distributions to draw up a space that allows us to better understand the outcome. For example, I suggest the following code:

```
num_points = 2000
vectors_set = []
for i in xrange(num_points):
 if np.random.random() > 0.5:
   vectors_set.append([np.random.normal(0.0, 0.9),
           np.random.normal(0.0, 0.9)])
 else:
   vectors_set.append([np.random.normal(3.0, 0.5),
           np.random.normal(1.0, 0.5)])
```

As we have done in the previous chapter, we can use some Python graphic libraries to plot the data. I propose that we use *matplotlib* like before, but this time we will also use the visualization package *Seaborn* based on *matplotlib* and the data manipulation package *pandas*, which allows us to work with more complex data structures.

If you do not have these packages installed, you must do it with the *pip* value before you can run the following codes.

To display the points that have been generated randomly I suggest the following code:

```
import matplotlib.pyplot as plt
import pandas as pd
import seaborn as sns

df = pd.DataFrame({"x": [v[0] for v in vectors_set],
          "y": [v[1] for v in vectors_set]})
sns.lmplot("x", "y", data=df, fit_reg=False, size=6)
plt.show()
```

This code generates a graph of points in a two dimensional space like the following screenshot:

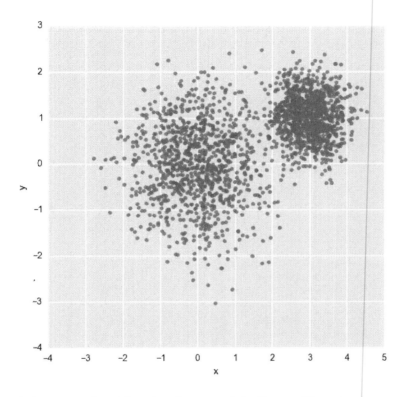

A *k-means* algorithm implemented in TensorFlow to group the above points, for example in four clusters, can be as follows (based on the model proposed by Shawn Simister in his blog[21]):

[21]   *Github*, (2016) Shawn Simister. [Online]. Available at: https://gist.github.com/narphorium/d06b7ed234287e319f18 [Accessed: 9/01/2016].

```python
import numpy as np

vectors = tf.constant(vectors_set)
k = 4
centroides =
tf.Variable(tf.slice(tf.random_shuffle(vectors),[0,0],[k,-1]))

expanded_vectors = tf.expand_dims(vectors, 0)
expanded_centroides = tf.expand_dims(centroides, 1)

assignments =
tf.argmin(tf.reduce_sum(tf.square(tf.sub(expanded_vectors,
                expanded_centroides)), 2), 0)

means = tf.concat(0, [tf.reduce_mean(tf.gather(vectors,
tf.reshape(tf.where( tf.equal(assignments, c)),[1,-1])), reduc-
tion_indices=[1]) for c in xrange(k)])

update_centroides = tf.assign(centroides, means)

init_op = tf.initialize_all_variables()

sess = tf.Session()
sess.run(init_op)

for step in xrange(100):
   _, centroid_values, assignment_values =
sess.run([update_centroides,
          centroides, assignments])
```

I suggest the reader checks the result in the *assignment_values* tensor with the following code, which generates a graph as above:

```
data = {"x": [], "y": [], "cluster": []}

for i in xrange(len(assignment_values)):
 data["x"].append(vectors_set[i][0])
 data["y"].append(vectors_set[i][1])
 data["cluster"].append(assignment_values[i])

df = pd.DataFrame(data)
sns.lmplot("x", "y", data=df,
      fit_reg=False, size=6,
      hue="cluster", legend=False)

plt.show()
```

The screenshot with the result of the execution of my code it is shown in the following figure:

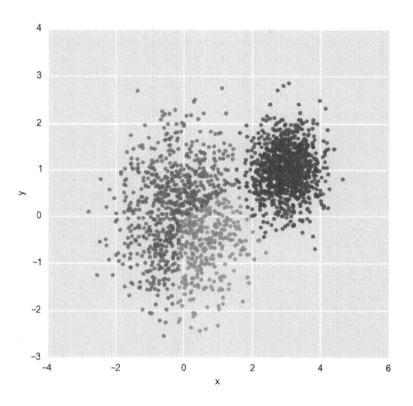

New groups

I assume that the reader might feel a little overwhelmed with the *K-means* code presented in the previous section. Well, I propose that we analyze this in detail, step by step, and especially watch the tensors invoved and how they are transformed during the program.

The first thing to do is move all our data to tensors. In a *constant tensor*, we keep our entry points randomly generated:

```
vectors = tf.constant(vectors_set)
```

Following the algorithm presented in the previous section, in order to start we must determine the initial centroids. As I advanced, an option may be randomly choose K observations from the input data. One way to do this is with the following code, which indicates to TensorFlow that it must shuffle randomly the entry point and choose the first K points as centroids:

```
k = 4
centroides =
tf.Variable(tf.slice(tf.random_shuffle(vectors),[0,0],[k,-1]))
```

These K points are stored in a 2D tensor. To know the shape of those tensors we can use *tf.Tensor.get_shape()*:

```
print vectors.get_shape()
print centroides.get_shape()

TensorShape([Dimension(2000), Dimension(2)])
TensorShape([Dimension(4), Dimension(2)])
```

We can see that *vectors* is an array that dimension D0 contains 2000 positions, one for each, and D1 contains the position *x,y* for each point. Instead, *centroids* is a matrix of four positions in the dimension D0, one position for each centroid, and the dimension D1 is equivalent to the dimension D1 of *vectors*.

Next, the algorithm enters in a loop. The first step is to calculate, for each point, its closest centroid by the *Squared Euclidean Distance*[22] (which can only be used when we want to compare distances):

$$d^2 \left(vector, centroide \right) = \left(vector_x - centroide_x \right)^2 + \left(vector_y - centroide_y \right)^2$$

To calculate this value *tf.sub(vectors, centroides)* is used. We should note that, although the two subtract tensors have both 2 dimensions, they have different sizes in one dimension (2000 vs 4 in dimension D0), which, in fact, also represent different things.

To fix this problem we could use some of the functions discussed before, for instance *tf.expand_dims* in order to insert a dimension in both tensors. The aim is to extend both tensors from 2 dimensions to 3 dimensions to make the sizes match in order to perform a subtraction:

```
expanded_vectors = tf.expand_dims(vectors, 0)
expanded_centroides = tf.expand_dims(centroides, 1)
```

---

[22] Wikipedia, (2016). Squared Euclidean distance. [Online]. Available at: https://en.wikipedia.org/wiki/Euclidean_distance# Squared_Euclidean_distance [Accessed: 9/01/2016].

tf.expand_dims inserts one dimension in each tensor; in the first dimension (D0) of *vectors* tensor, and in the second dimension (D1) of *centroids* tensor. Graphically, we can see that in the extended tensors the dimensions have the same meaning in each of them:

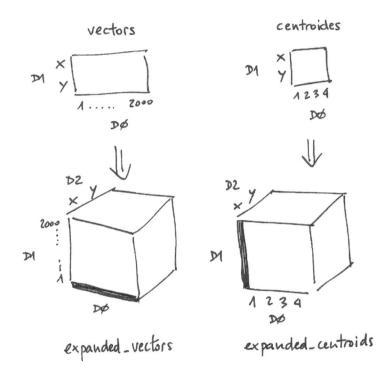

66

It seems to be solved, but actually, if you look closely (out-lined in bulk in the illustration), in each case there are dimensions that have not been able to determinate the sizes of those dimensions. Remember that the with *get_shape()* function we can find out:

```
print expanded_vectors.get_shape()
print expanded_centroides.get_shape()
```

The output is as follows:

```
TensorShape([Dimension(1), Dimension(2000), Dimension(2)])
TensorShape([Dimension(4), Dimension(1), Dimension(2)])
```

With 1 it is indicating a no assigned size.

But I have already advanced that TensorFlow allows *broadcasting*, and therefore the *tf.sub* function is able to discover for itself how to do the subtraction of elements between the two tensors.

Intuitively, and observing the previous drawings, we see that the shape of the two tensors match, and in those cases both tensors have the same size in a certain dimension. These math, as happens in dimension D2. Instead, in the dimension D0 only has a defined size the *expanded_centroides*.

In this case, TensorFlow assumes that the dimension D0 of *expanded_vectors* tensor have to be the same size if we want to perform a subtraction element to element within this dimension.

And the same happens with the size of the dimension D1 of *expended_centroides* tensor, where TensorFlow deduces the size of the dimension D1 of *expanded_vectors* tensor.

Therefore, in the allocation step (step 1) the algorithm can be expressed in these four lines of TensorFlow's code, which calculates the *Squared Euclidean Distance*:

```
diff=tf.sub(expanded_vectors, expanded_centroides)
sqr= tf.square(diff)
distances = tf.reduce_sum(sqr, 2)
assignments = tf.argmin(distances, 0)
```

And, if we look at the shapes of tensors, we see that they are respectively for *diff, sqr, distances* and *assignments* as follows:

```
TensorShape([Dimension(4), Dimension(2000), Dimension(2)])
TensorShape([Dimension(4), Dimension(2000), Dimension(2)])
TensorShape([Dimension(4), Dimension(2000)])
TensorShape([Dimension(2000)])
```

That is, *tf.sub* function has returned the tensor *dist*, that contains the subtraction of the index values for centroids and vector (indicated in the dimension D1, and the centroid indicated in the dimension D0. For each index *x,y* are indicated in the dimension D2) .

The *sqr* tensor contains the square of those. In the *distance* tensor we can see that it has already reduced one dimension, the one indicated as a parameter in *tf.reduce_sum* function.

I use this example to explain that TensorFlow provides several operations which can be used to perform mathematical operations that reduce dimensions of a tensor as in the case of *tf.reduce_sum*. In the table below you can find a summary of the most important ones:

Operation	Description
tf.reduce_sum	Computes the sum of the elements along one dimension
tf.reduce_prod	Computes the product of the elements along one dimension
tf.reduce_min	Computes the minimum of the elements along one dimension
tf.reduce_max	Computes the maximum of the elements along one dimension
tf.reduce_mean	Computes the mean of the elements along one dimension

Finally, the assignation is achieved with *tf.argmin*, which returns the index with the minimum value of the tensor *dimension* (in our case $D0$, which remember that was the centroid). We also have the *tf.argmax* operation:

Operation	Description
tf.argmin	Returns the index of the element with the minimum value along *tensor* dimension
tf.argmax	Returns the index of the element with the maximum value of the *tensor* dimension

In fact, the 4 instructions seen above could be summarized in only one code line, as we have seen in the previous section:

```
assignments =
tf.argmin(tf.reduce_sum(tf.square(tf.sub(expanded_vectors,
            expanded_centroides)), 2), 0)
```

But anyway, internal *tensors* and the operations that they define as nodes and execute the internal graph are like the ones we have described before.

Computation of the new centroids

Once we have created new groups on each iteration, we will have to remember that the new step of the algorithm consists in calculating the new centroids of the groups. In the code of the section before we have seen this line of code:

```
means = tf.concat(0, [tf.reduce_mean(tf.gather(vectors,
tf.reshape(tf.where( tf.equal(assignments, c)),[1,-1])), reduc-
tion_indices=[1]) for c in xrange(k)])
```

On that piece of code, we can see that the *means* tensor is the result of the concatenation of the k tensors that correspond to the mean value of every point that belongs to each k *cluster*.

Next, I will comment on each of the TensorFlow operations that are involved in the calculation of the mean value of every points that belongs to each *cluster*[23]:

- With *tf.equal* we can obtain a *boolean tensor (Dimension(2000))* that indicates (with *true* value) the positions where the *assignment tensor* match with the K *cluster*, which, at the time, we are calculating the average value of the points.

- With *tf.where* is constructed a *tensor (Dimension(1) x Dimension(2000))* with the position where the values *true* are on the *boolean tensor* received as a parameter. i.e. a list of the position of these.

- With *tf.reshape* is constructed a tensor *(Dimension(2000) x Dimension(1))* with the index of the points inside *vectors* tensor that belongs to this *c cluster*.

- With *tf.gather* is constructed a *tensor (Dimension(1) x Dimension(2000))* which gathers the coordenates of the points that form the *c cluster*.

- With *tf.reduce_mean* it is constructed a tensor *(Dimension(1) x Dimension(2))* that contains the average value of all points that belongs to the cluster *c*.

---

[23] *In my opinion, the level of explanation of each operation it's enough for the purpose of this book.*

Anyway, if the reader wants to dig deeper into the code, as I always say, you can find more info for each of these operations, with very illustrative examples, on the *TensorFlow API* page[24].

Graph Execution

Finally, we have to describe the part of the above code that corresponds to the loop and to the part that update the centroids with the new values of the *means* tensor.

To do this, we need to create an operator that assigns the value of the variable *means* tensor into *centroids* in a way than, when the operation *run()* is executed, the values of the updated centroids are used in the next iteration of the loop:

```
update_centroides = tf.assign(centroides, means)
```

We also have to create an operator to initialize all of the variable before starting to run the graph:

```
init_op = tf.initialize_all_variables()
```

---

[24] TensorFlow, (2016) Python API. [online]. Available in: https://www.tensorflow.org/versions/master/api_docs/index.html [Accessed: 19/02/2016].

At this point everything is ready. We can start running the graph:

```
sess = tf.Session()
sess.run(init_op)

for step in xrange(num_steps):
  _, centroid_values, assignment_values =
sess.run([update_centroides,
          centroides,
          assignments])
```

In this code, for each iteration, the centroids and the new allocation of clusters for each entry points are updated.

Notice that the code specifies three operators and it has to go look in the execution of the call *run()*, and running in this order. Since there are three values to search, *sess.run()* returns a data structure of three *numpy array* elements with the contents of the corresponding tensor during the training process.

As *update_centroides* is an operation whose result is not the parameter that returns, the corresponding item in the return tuple contains nothing, and therefore be ruled out, indicating it with "_"[25].

For the other two values, the centroids and the assigning points to each cluster, we are interested in displaying them on screen once they have completed all *num_steps* iterations.

---

[25] Actually "_" is like any other variable, but many Python users, by convention, we use it to discard results.

We can use a simple print:

```
print centroid_values
```

The output is as it follows:

```
[[ 2.99835277e+00 9.89548564e-01]
 [ -8.30736756e-01 4.07433510e-01]
 [ 7.49640584e-01 4.99431938e-01]
 [ 1.83571398e-03 -9.78474259e-01]]
```

I hope that the reader has a similar values on the screen, since this will indicate that he has successfully executed the proposed code in this chapter of the book.

I suggest that the reader tries to change any of the values in the code, before advancing. For example the num_points, and especially the number of clustersk, and see how it changes the result in the *assignment_values* tensor with the previous code that generates a graph.

Remember that in order to facilitate testing the code described in this chapter, it can be downloaded from Github[26] . The name of the file that contains this code is *Kmeans.py*.

----

[26]   Github, (2016) TensorFlow Book – Jordi Torres. [online]. Available at:
https://github.com/jorditorresBCN/LibroTensorFlow
[Accessed: 19/02/2016].

In this chapter we have advanced some knowledge of TensorFlow, especially on basic data structure *tensor*, from a code example in TensorFlow that implements a clustering algorithm *K-means*.

With this knowledge, we are ready to build a single layer neural network, step by step, with TensorFlow in the next chapter.

4. Single Layer Neural Network in TensorFlow

In the preface, I commented that one of the usual uses of Deep Learning includes pattern recognition. Given that, in the same way that beginners learn a programming language starting by printing "Hello World" on screen, in Deep Learning we start by recognizing hand-written numbers.

In this chapter I present how to build, step by step, a neural network with a single layer in TensorFlow. This neural network will recognize hand-written digits, and it's based in one of the diverse examples of the beginner's tutorial of Tensor-Flow[27].

Given the introductory style of this book, I chose to guide the reader while simplifying some concepts and theoretical justifications at some steps through the example.

If the reader is interested in learn more about the theoretical concepts of this example after reading this chapter, I suggest

---

[27] TensorFlow, (2016) Tutorial MNIST beginners. [online]. Available at: https://www.tensorflow.org/versions/master/tutorials/mnist/beginners [Accessed: 16/12/2015].

to read *Neural Networks and Deep Learning*[28], available online, presenting this example but going in depth with the theoretical concepts.

The MNIST Data-set

The MNIST data-set is composed by a set of black and white images containing hand-written digits, containing more than 60.000 examples for training a model, and 10.000 for testing it. The MNIST data-set can be found at the *MNIST database*[29].

This data-set is ideal for most of the people who begin with pattern recognition on real examples without having to spend time on data pre-processing or formatting, two very important steps when dealing with images but expensive in time.

The black and white images (bilevel) have been normalized into 20x20 pixel images, preserving the aspect ratio. For this case, we notice that the images contain gray pixels as a result

---

[28]   Neural Networks and Deep Learning. Michael Nielsen. [online]. Available at:
http://neuralnetworksanddeeplearning.com/index.html
[Accessed: 6/12/2015].
[29]   The MNIST database of handwritten digits.[online]. Available at: http://yann.lecun.com/exdb/mnist
[Accessed: 16/12/2015].

of the *anti-aliasing*[30] used in the normalization algorithm (reducing the resolution of all the images to one of the lowest levels). After that, the images are centered in 28x28 pixel frames by computing the mass center and moving it into the center of the frame. The images are like the ones shown here:

Also, the kind of learning required for this example is *supervised learning*; the images are labeled with the digit they represent. This is the most common form of Machine Learning.

In this case we first collect a large data set of images of numbers, each labelled with its value. During the training, the model is shown an image and produces an output in the form of a vector of scores, one score for each category. We want the desired category to have the highest score of all categories, but this is unlikely to happen before training.

We compute an objective function that measures the error (as we did in previous chapters) between the output scores and the desired pattern of scores. The model then modifies its internal adjustable parameters , called weights, to reduce this error. In a typical Deep Learning system, there may be hundreds of millions of these adjustable weights, and

---

[30] Wikipedia, (2016). Antialiasing [online]. Available at:
https://en.wikipedia.org/wiki/Antialiasing [Accessed: 9/01/2016].

hundreds of millions of labelled examples with which to train the machine. We will consider a smaller example in order to help the understanding of how this type of models work.

To download easily the data, you can use the script *input_data.py*[31], obtained from Google's site[32] but uploaded to the book's *github* for your comodity. Simply download the code *input_data.py* in the same work directory where you are programming the neural network with TensorFlow. From your application you only need to import and use in the following way:

```
import input_data
mnist = input_data.read_data_sets("MNIST_data/", one_hot=True)
```

After executing these two instructions you will have the full training data-set in *mnist.train* and the test data-set in *mnist.test*. As I previously said, each element is composed by an image, referenced as *"xs"*, and its corresponding label *"ys"*, to make easier to express the processing code. Remember that all data-sets, training and testing, contain *"xs"* and

---

[31]   *Github*, (2016) Book TensorFlow – Jordi Torres. [online].
Available at: https://github.com/jorditorresBCN/
LibroTensorFlow/blob/master/input_data.py
[Accessed: 9/01/2016].
[32]   Google (2016) TensorFlow. [online]. Available at:
https://tensorflow.googlesource.com [Accessed: 9/01/2016].

"*ys*"; also, the training images are referenced in *mnist.train.images* and the training labels in *mnist.train.labels*.

As previously explained, the images are formed by 28x28 pixels, and can be represented as a numerical matix. For example, one of the images of number 1 can be represented as:

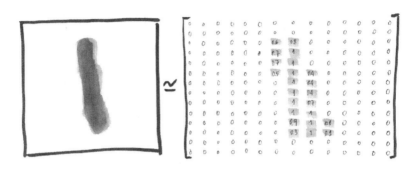

Where each position indicates the level of lackness of each pixel between 0 and 1. This matrix can be represented as an array of 28×28 = 784 numbers. Actually, the image has been transformed in a bunch of points in a vectorial space of 784 dimensions. Only to mention that when we reduce the structure to 2 dimensions, we can be losing part of the information, and for some computer vision algorithms this could affect their result, but for the simplest method used in this tutorial this will not be a problem.

Summarizing, we have a *tensor mnist.train.images* in 2D in which calling the fuction *get_shape()* indicates its shape:

TensorShape([Dimension(60000), Dimension(784)])

The first dimension indexes each image and the second each pixel in each image. Each element of the tensor is the intensity of each pixel between 0 and 1.

Also, we have the labels in the form of numbers between 0 and 9, indicating which digit each image represents. In this example, we are representing labels as a vector of 10 positions, which the corresponding position for the represented number contains a 1 and the rest 0. So *mnist.train.labels*es is a *tensor* shaped as *TensorShape([Dimension(60000), Dimension10)])*.

An artificial neuron

Although the book doesn't focus on the theoretical concepts of neural netwoks, a brief and intuitive introduction of how neurons work to learn the training data will help the reader to undertand what is happening. Those readers that already know the theory and just seek how to use TensorFlow can skip this section.

Let's see a simple but illustrative example of how a neuron learns. Suppose a set of points in a plane labeled as "square" and "circle". Given a new point "X", we want to know which label corresponds to it:

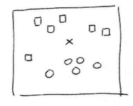

A usual approximation could be to draw a straight line dividing the two groups and use it as a classifier:

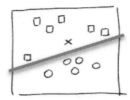

In this situation, the input data is represented by vectors shaped as (x,y) representing the coordinates in this 2-dimension space, and our function returning '0' or '1' (above or below the line) to know how to classify it as a "square" or "circle". Mathematically, as we learned in the linear regression chapter, the "line" (classifier) can be expressed as $y= W*x+b$.

Generalizing, a neuron must learn a weight W (with the same dimension as the input data X) and an offset b (called *bias* in neural networks) to learn how to classify those values. With them, the neuron will compute a weighted sum of the inputs in X using weight W, and add the offset b; and finally the neuron will apply an "activation" non-linear function to produce the result of "0" or "1".

The function of the neuron can be more formally expressed as:

$$z = b + \sum_i x_i w_i$$

$$y = \begin{cases} 1 & \text{si } z \geqslant 0 \\ 0 & \text{si } z < 0 \end{cases}$$

Having defined this function for our neuron, we want to know how the neuron can learn those parameters W and b from the labeled data with "squares" and "circles" in our example, to later label the new point "X".

A first approach could be similar to what we did with the linear regression, this is, feed the neuron with the known labeled data, and compare the obtained result with the real one. Then, when iterating, weights in W and b are adjusted to minimize the error, as shown again in chapter 2 for the linear regression line.

Once we have the W and b parameters we can compute the weighted sum, and now we need the function to turn the result stored in z into a '0' or '1'. There are several activation functions available, and for this example we can use a popular one called *sigmoid*[33], returning a real value between 0 and 1:

$$z = b + \sum_i x_i w_i \qquad y = \frac{1}{1 + e^{-z}}$$

Looking at the formula we see that it will tend to return values close to 0 or 1. If input z is big enough and positive, "e" powered to minus z is zero and then y is 1. If the input z is big enough and negative, "e" powered to a large positive

[33] Wikipedia, (2016). Sigmoid function [online]. Avaliable at: https://en.wikipedia.org/wiki/Sigmoid_function [Accessed: 12/01/2016].

number becomes also a large positive number, so the denominator becomes large and the final y becomes 0. If we plot the function it would look like this:

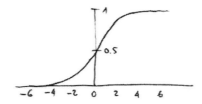

Since here we have presented how to define a neuron, but a neural network is actually a composition of neurons connected among them in a different ways and using different activation functions. Given the scope of this book, I'll not enter into all the extension of the neural networks universe, but I assure you that it is really exciting.

Just to mention that there is a specific case of neural networks (in which Chapter 5 is based on) where the neurons are organized in layers, in a way where the inferior layer (input layer) receives the inputs, and the top layer (output layer) produces the response values. The neural network can have several intermediate layers, called hidden layers. A visual way to represent this is:

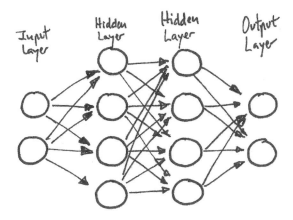

In these networks, the neurons of a layer communicate to the neurons of the previous layer to receive information, and then communicate their results to the neurons of the next layer.

As previously said, there are more activation functions apart of the *Sigmoid*, each one with different properties. For example, when we want to classify data into more than two classes at the output layer, we can use the *Softmax*[34] activation function, a generalization of the *sigmoid* function. *Softmax* allows obtaining the probability of each class, so their sum is 1

[34] Wikipedia, (2016). Softmax function [online]. Available at: https://en.wikipedia.org/wiki/Softmax_function [Accessed: 2/01/2016].

and the most probable result is the one with higher probability.

An easy example to start: Softmax

Remember that the problem to solve is that, given an input image, we get the probability that it belongs to a certain digit. For example, our model could predict a "9" in an image with an 80% certainty, but give a 5% of chances to be an "8" (due to a dubious lower trace), and also give certain low probabilities to be any other number. There is some uncertainty on recognizing hand-written numbers, and we can't recognize the digits with a 100% of confidence. In this case, a probability distribution gives us a better idea of how much confidence we have in our prediction.

So, we have an output vector with the probability distribution for the different output labels, mutually exclusive. This is, a vector with 10 probability values, each one corresponding to each digit from 0 to 9, and all probabilities summing 1.

As previously said, we get to this by using an output layer with the *softmax* activation function. The output of a neuron with a *softmax* function depends on the output of the other neurons of its layer, as all of their outputs must sum 1.

The *softmax* function has two main steps: first, the "evidences" for an image belonging to a certain label are computed, and later the evidences are converted into probabilities for each possible label.

Evidence of belonging

Measuring the evidence of a certain image to belong to a specific class/label, a usual approximation is to compute the weighted sum of pixel intensities. That weight is negative when a pixel with high intensity happens to not to be in a given class, and positive if the pixel is frequent in that class.

Let's show a graphical example: suppose a learned model for the digit "0" (we will see how this is learned later). At this time, we define a model as "something" that contains information to know whether a number belongs to a specific class. In this case, we chose a model like the one below, where the red (or bright gray for the b/n edition) represents negative examples (this is, reduce the support for those pixels present in "0"), while the blue (the darker gray for b/n edition) represents the positive examples. Look at it:

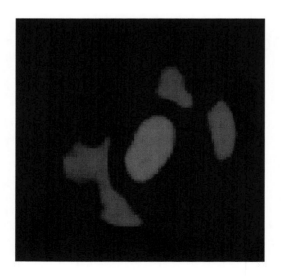

Think in a white paper sheet of 28x28 pixels and draw a "0" on it. Generally our zero would be drawn in the blue zone (remember that we left some space around the 20x20 drawing zone, centering it later).

It is clear that in those cases where our drawing crosses the red zone, it is most probably that we are not drawing a zero. So, using a metric based in rewarding those pixels, stepping on the blue region and penalizing those stepping the red zone seems reasonable.

Now think of a "3": it is clear that the red zone of our model for "0" will penalize the probabilities of it being a "0". But if the reference model is the following one, generally the pixels forming the "3" will follow the blue region; also the drawing of a "0" would step into a red zone.

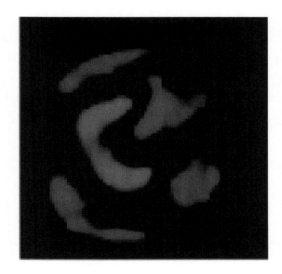

I hope that the reader, seeing those two examples, understands how the explained approximation allows us to estimate which number represent those drawings.

The following figure shows an example of the ten different labels/classes learned from the MNIST data-set (extracted from the examples of Tensorflow[35]). Remember that red (bright gray) represents negative weights and blue (dark gray) represents the positive ones.

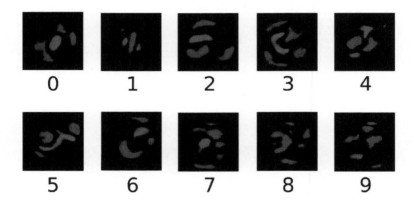

In a more formal way, we can say that the evidence for a class i given an input x is expressed as:

$$evidence_i = \sum_j W_{i,j} x_j$$

---

[35] TensorFlow, (2016) Tutorial MNIST beginners. [online].
Available at: https://www.tensorflow.org/versions/master/
tutorials/mnist/beginners [Accessed: 16/12/2015].

Where i indicates the class (in our situation, between 0 and 9), and j is an index to sum the indexes of our input image. Finally Wi represents the aforementioned weights.

Remember that, in general, the models also include an extra parameter representing the bias, adding some base uncertainty. In our situation, the formula would end like this:

$$evidence_i = \sum_j W_{ij} x_j + b_i$$

For each i (between 0 and 9) we have a matrix Wi of 784 elements (28x28), where each element j is multiplied by the corresponding component j of the input image, with 784 components, then added b_i. A graphical view of the matrix calculus and indexes is this:

Probability of belonging

We commented that the second step consisted on computing probabilities. Specifically we turn the sum of evidences into predicted probabilities, indicated as y, using the *softmax* function:

$$y = softmax\,(evidence)$$

Remember that the output vector must be a probability function with sum equals 1. To normalize each component, the softmax function uses the exponential value of each of its inputs and then normalizes them as follows:

$$softmax\,(x)_i = \frac{exp(x_i)}{\sum_j exp(x_j)} = \frac{e^{x_i}}{\sum_j e^{x_j}}$$

The obtained effect when using exponentials is a multiplication effect in weights. Also, when the evidence for a class is small, this class support is reduced by a fraction of its previous weight. Furthermore, *softmax* normalizes the weights making them to sum 1, creating a probability distribution.

The interesting fact of such function is that a good prediction will have one output with a value near 1, while all the other outputs will be near zero; and in a *weak* prediction, some labels may show similar support.

Programming in TensorFlow

After this brief description of what the algorithm does to recognize digits, we can implement it in TensorFlow. For this, we can make a quick look at how the *tensors* should be to store our data and the parameters for our model. For this purpose, the following schema depicts the data structures and their relations (to help the reader recall easily each piece of our problem):

First of all we create two variables to contain the weights W and the bias b:

```
W = tf.Variable(tf.zeros([784,10]))
b = tf.Variable(tf.zeros([10]))
```

Those variables are created using the *tf.Variable* function and the initial value for the variables; in this case we initialize the tensor with a constant tensor containing zeros.

We see that W is shaped as *[Dimension(784), Dimension(10)]*, defined by its argument, a constant tensor *tf.zeros[784,10]* dimensioned like *W*. The same happens with the bias *b*, shaped by its argument as *[Dimension(10)]*.

Matrix W has that size because we want to multiply the image vector of 784 positions for each one of the 10 possible digits, and produce a tensor of evidences after adding *b*.

In this case of study using MNIST, we also create a tensor of two dimensions to keep the information of the *x* points, with the following line of code:

```
x = tf.placeholder("float", [None, 784])
```

The tensor *x* will be used to store the MNIST images as a vector of 784 floating point values (using *None* we indicate that the dimension can be any size; in our case it will be equal to the number of elements included in the learning process).

Now we have the tensors defined, and we can implement our model. For this, TensorFlow provides several operations, being *tf.nn.softmax(logits, name=None)* one of the available ones, implementing the previously described *softmax* function. The arguments must be a tensor, and optionally, a name. The function returns a tensor of the same kind and shape of the tensor passed as argument.

In our case, we provide to this function the resulting tensor of multiplying the image vector x and the weight matrix W, adding b:

```
y = tf.nn.softmax(tf.matmul(x,W) + b)
```

Once specified the model implementation, we can specify the necessary code to obtain the weights for W and bias b using an iterative training algorithm. For each iteration, the training algorithm gets the training data, applies the neural network and compares the obtained result with the expected one.

To decide when a model is good enough or not we have to define what "good enough" means. As seen in prevous chapters, the usual methodology is to define the opposite: how "bad" a model is using *cost* functions. In this case, the goal is to obtain values of W and b that minimizes the function indicating how "bad" the model is.

There are different metrics for the degree of error between resulting outputs and expected outputs on training data. One common metric is the *mean squared error* or the *squared Euclidean distance*, which have been previously seen. Even though, some research lines propose other metrics for such purpose in neural networks, like the *cross entropy error*, used in our example. This metric is computed like this:

$$-\sum_i y_i' \log(y_i)$$

Where y is the predicted distribution of probability and y' is the real distribution, obtained from the labeling of the training data-set. We will not enter into details of the maths behind the cross-entropy and its place in neural networks, as it is far more complex than the intended scope of this book; just indicate that the minimum value is obtained when the two distributions are the same. Again, if the reader wants to learn the insights of this function, we recommend reading *Neural Networks and Deep Learning*[36].

To implement the *cross-entropy* measurement we need a new placeholder for the correct labels:

```
y_ = tf.placeholder("float", [None,10])
```

Using this placeholder, we can implement the *cross-entropy* with the following line of code, representing our cost function:

```
cross_entropy = -tf.reduce_sum(y_*tf.log(y))
```

First, we calculate the logarithm of each element y with the built-in function in TensorFlow *tf.log()*, and then we multiply it for each $y\_$ element. Finally, using *tf.reduce_sum* we sum all the elements of the tensor (later we will see that the images are visited in bundles, and in this case the value of *cross-*

[36]   Neural Networks & Deep Learning.Michael Nielsen. [online]. Available at: http://neuralnetworksanddeeplearning.com/ index.html [Accessed: 6/12/2015].

entropy corresponds to the bundle of images *y* and not a single image).

Iteratively, once determined the error for a sample, we have to correct the model (modifying the parameters *W* and *b*, in our case) to reduce the difference between computed and expected outputs in the next iteration.

Finally, it only remains to specify this iterative minimization process. There are several algorithms for this purpose in neural networks; we will use the *backpropagation* (backward propagation of errors) algorithm, and as its name indicates, it propagates backwards the error obtained at the outputs to recompute the weights of *W*, especially important for multi-layer neural network.

This method is used together with the previously seen gradient descent method, which using the cross-entropy cost function allows us to compute how much the parameters must change on each iteration in order to reduce the error using the available local information at each moment. In our case, intuitively it consist on changing the weights *W* a little bit on each iteration (this little bit expressed by a *learning rate* hyperparameter, indicating the speed of change) to reduce the error.

Due that in our case we only have one layer neural network we will not enter into backpropagation methods. Only remember that TensorFlow knows the entire computing graph, allowing it to apply optimization algorithms to find the proper gradients for the cost function to train the model.

So, in our example using the MNIST images, the following line of code indicates that we are using the backpropagation algorithm to minimize the *cross-entropy* using the *gradient descent* algorithm and a *learning rate* of 0.01:

```
train_step =
tf.train.GradientDescentOptimizer(0.01).minimize(cross_entropy)
```

Once here, we have specified all the problem and we can start the computation by instantiating *tf.Session()* in charge of executing the TensorFlow operations in the available devices on the system, CPUs or GPUs:

```
sess = tf.Session()
```

Next, we can execute the operation initializing all the variables:

```
sess.run(tf.initialize_all_variables())
```

From this moment on, we can start training our model. The returning parameter for *train_step*, when executed, will apply the gradient descent to the involved parameters. So training the model can be achieved by repeating the *train_step* execution. Let's say that we want to iterate 1.000 times our *train_step*; we have to indicate the following lines of code:

```
for i in range(1000):
  batch_xs, batch_ys = mnist.train.next_batch(100)
  sess.run(train_step, feed_dict={x: batch_xs, y_: batch_ys})
```

The first line inside the loop specifies that, for each iteration, a bundle of 100 inputs of data, randomly sampled from the training data-set, are picked. We could use all the training data on each iteration, but in order to make this first example more agile we are using a small sample each time. The second line indicates that the previously obtained inputs must feed the respective placeholders.

Finally, just mention that the Machine Learning algorithms based in gradient descent can take advantage from the capabilities of automatic differentiation of TensorFlow. A TensorFlow user only has to define the computational architecture of the predictive model, combine it with the target function, and then just add the data.

TensorFlow already manages the associated calculus of derivatives behind the learning process. When the *minimize()* method is executed, TensorFlow identifies the set of variables on which *loss function* depends, and computes gradients for each of these. If you are interested to know how the differentiation is implemented you can inspect the *ops/gradients.py* file[37].

---

[37] TensorFlow Github: tensorflow/tensorflow/python/ops/ gradients.py [Online]. Available at:

https://github.com/tensorflow/tensorflow/blob/master/tensorflow /python/ops/gradients.py [Accessed: 16/03/2016].

Model evaluation

A model must be evaluated after training to see how much "good" (or "bad") is. For example, we can compute the percentage of hits and misses in our prediction, seeing which examples were correctly predicted. In previous chapters, we saw that the *tf.argmax(y,1)* function returns the index of the highest value of a tensor according a given axis. In effect, *tf.argmax(y,1)* is the label in our model with higher probability for each input, while *tf.argmax(y_,1)* is the correct label. Using *tf.equal* method we can compare if our prediction coincides with the correct label:

```
correct_prediction = tf.equal(tf.argmax(y,1), tf.argmax(y_,1))
```

This instruction returns a list of Booleans. To determine which fractions of predictions are correct, we can cast the values to numeric variables (floating point) and do the following operation:

```
accuracy = tf.reduce_mean(tf.cast(correct_prediction, "float"))
```

For example, *[True, False, True, True]* will turn into *[1,0,1,1]* and the average will be 0.75 representing the percentage of *accuracy*. Now we can ask for the accuracy of our test data-set using the *mnist.test* as the *feed_dict* argument:

```
print sess.run(accuracy, feed_dict={x: mnist.test.images, y_:
mnist.test.labels})
```

I obtained a value around 91%. Are these results good? I think that they are fantastic, because this means that the reader has been able to program and execute her first neural network using TensorFlow.

Another problem is that other models may provide better accuracy, and this will be presented in the next chapter with a neural network containing more layers.

The reader will find the whole code used in this chapter in the file *RedNeuronalSimple.py*, in the book's *github*[38]. Just to provide a global view of it, I'll put it here together:

```
import input_data
mnist = input_data.read_data_sets("MNIST_data/", one_hot=True)

import tensorflow as tf

x = tf.placeholder("float", [None, 784])
W = tf.Variable(tf.zeros([784,10]))
b = tf.Variable(tf.zeros([10]))

matm=tf.matmul(x,W)
y = tf.nn.softmax(tf.matmul(x,W) + b)
y_ = tf.placeholder("float", [None,10])

cross_entropy = -tf.reduce_sum(y_*tf.log(y))
train_step =
tf.train.GradientDescentOptimizer(0.01).minimize(cross_entropy)

sess = tf.Session()
sess.run(tf.initialize_all_variables())
```

---

[38] *Github*, (2016) Libro TensorFlow – Jordi Torres. [online]. Available at: https://github.com/jorditorresBCN/LibroTensorFlow [Accessed: 9/01/2016].

```
for i in range(1000):
  batch_xs, batch_ys = mnist.train.next_batch(100)
  sess.run(train_step, feed_dict={x: batch_xs, y_: batch_ys})
  correct_prediction = tf.equal(tf.argmax(y,1), tf.argmax(y_,1))
  accuracy = tf.reduce_mean(tf.cast(correct_prediction, "float"))
  print sess.run(accuracy, feed_dict={x: mnist.test.images, y_:
mnist.test.labels})
```

5. Multi-layer Neural Networks in TensorFlow

In this chapter I will program, with the reader, a simple Deep Learning neural network using the same MNIST digit recognition problem of the previous chapter.

As I have advanced, a Deep Learning neural network consists of several layers stacked on top of each other. Specifically, in this chapter we will build a convolutional network, this is, an archetypal example of Deep Learning. Convolution neural networks were introduced and popularized in 1998 by Yann LeCunn and others. These convolutional networks have recently led state of the art performance in image recognition; for example: in our case of digit recognition they achieve an accuracy higher than 99%.

In the rest of this chapter, I will use an example code as the backbone, alongside which I will explain the two most important concepts of these networks: convolutions and pooling without entering in the details of the parameters, given the introductory nature of this book. However, the reader will be able to run all the code and I hope that it will allow you understand to global ideas behind convolutional networks.

Convolutional Neural Networks

Convolutional Neural Nets (also known as CNN's or Cov-Nets) are a particular case of Deep Learning and have had a significant impact in the area of computer vision.

A typical feature of CNN's is that they nearly always have images as inputs, this allows for more efficient implementation and a reduction in the number of required parameters. Let's have a look at our MNIST digit recognition example: after reading in the MNIST data and defining the placeholders using TensorFlow as we did in the previous example:

```
import input_data
mnist = input_data.read_data_sets('MNIST_data', one_hot=True)

import tensorflow as tf

x = tf.placeholder("float", shape=[None, 784])
y_ = tf.placeholder("float", shape=[None, 10])
```

We can reconstruct the original shape of the images of the input data. We can do this as follows:

```
x_image = tf.reshape(x, [-1,28,28,1])
```

Here we changed the input shape to a 4D tensor, the second and third dimension correspond to the width and the height of the image while the last dimension corresponding number of color channels, 1 in this case.

This way we can think of the input to our neural network be-ing a 2 dimensional space of neurons with size of 28x28 as depicted in the figure:

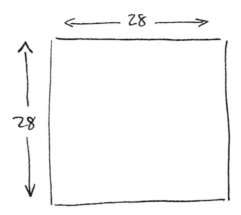

There are two basic principles that define convolution neural networks: the filters and the characteristic maps. These principles can be expressed as groups of specialized neurons, as we will see shortly. But first, we will give a short description of these two principles given their importance in CNN's.

Intuitively, we could say that the main purpose of a convolutional layer is to detect characteristics or visual features in the images, think of edges, lines, blobs of color, etc. This is taken care of by a hidden layer that is connected by to input layer that we just discussed. In the case of CNN's, in which we are interested, the input data is not fully connected to the neurons of the first hidden layer; this only happens in a small localized space in the input neurons that store the pixels values of the image. This can be visualized as follows:

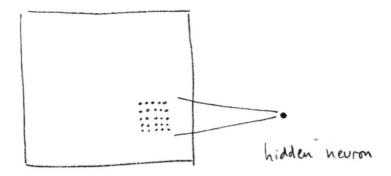

hidden neuron

To be more precise, in the given example each neuron of our hidden layer is connected with a small 5x5 region (hence 25 neurons) of the input layer.

We can think of this being a window of size 5x5 that passes over the entire input layer of size 28x28 that contains the input image. The window slides over the entire layer of neurons. For each position of the window there is a neuron in the hidden layer that processes that information.

We can visualize this by assuming that the window starts in the top left corner of the image; this provides the information to the first neuron of the hidden layer. The window is then slid right by one pixel; we connect this 5x5 region with the second neuron in the hidden layer. We continue like this until the entire space from top to bottom and from left to right has been convered by the window.

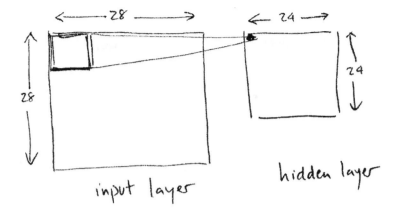

input layer

hidden layer

Analyzing the concrete case that we have proposed, we observe that given an input image of size 28x28 and a window of size 5x5 leads to a 24x24 space of neurons in the first hidden layer due to the fact we can only move the window 23 times down and 23 times to the right before hitting the bottom right edge of the input image. This assumes that window is moved just 1 pixel each time, so the new window overlaps with the old one expect in line that has just advanced.

It is, however, possible to move more than 1 pixel at a time in a convlution layer, this parameter is called the 'stride' length. Another extension is to pad the edges with zeroes (or other values) so that the window can slide over the edge of the image, which may lead to better results. The parameter to control this feature is know as padding,[39] with which you can determine the size of the padding. Given the introductory nature of the book we will not go into further detail about these two parameters.

Given our case of study, and following the formalism of the previous chapter, we will need a bias value b and a 5x5 weight matrix W to connect the neurons of the hidden layer with the input layer. A key feature of a CNN is that this weight matrix W and bias b are shared between all the neurons in the hidden layer; we use the same W and b for neurons in the hidden layer. In our case that is 24x24 (576) neurons. The reader should be able to see that this drastically reduces the amount weight parameters that one needs when compared to a fully connected neural network. To be specific, this reduces from 14000 (5x5x24x24) to just 25 (5x5) due to the sharing of the weight matrix W.

This shared matrix W and the bias b are usually called a kernel or filter in the context of CNN's. These filters are similar to those used image processing programs for retouching images, which in our case are used to find discriminating features. I recommend looking at the examples found in the GIMP[40] manual to get a good idea on how process of convolution works.

---

[39] The reader can read more about thre details of these parameters on the course website of CS231 - *Convolutional Neural Networks for Visual Recognition* (2015) [online]. Available at: http://cs231n.github.io/convolutional-networks [Accessed: 30/12/2015].

[40] GIMP – *Image processing software by GNU*, Convlution matrix documentation available at: https://docs.gimp.org/es/plug-in-convmatrix.html [Accessed: 5/1/2016].

A matrix and a bias define a kernel. A kernel only detects one certain relevant feature in the image so it is, therefore, recommended to use several kernels, one for each characteristic we would like to detect. This means that a full convolution layer in a CNN consists of several kernels. The usual way of representing several kernels is as follows:

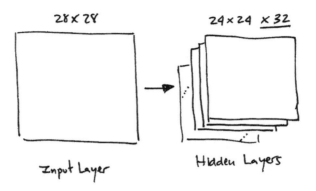

The first hidden layer is composed by several kernels. In our example, we use 32 kernels, each one defined by a 5x5 weight matrix W and a bias b that is also shared between the neuros of the hidden layer.

In order to simplify the code, I define the following two functions related to the weight matrix W and bias b:

```
def weight_variable(shape):
 initial = tf.truncated_normal(shape, stddev=0.1)
 return tf.Variable(initial)

def bias_variable(shape):
 initial = tf.constant(0.1, shape=shape)
 return tf.Variable(initial)
```

Without going into the details, it is customary to initialize the weights with some random noise and the bias values slightly positive.

In addition to the convolution layers that we just described, it is usual for the convolution layer to be followed by a so called pooling layer. The pooling layer simply condenses the output from the convolutional layer and creates a compact version of the information that have been put out by the convolutional layer. In our example, we will use a 2x2 region of the convolution layer of which we summarize the data into a single point using pooling:

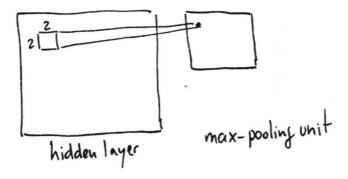

hidden layer   max-pooling unit

There are several ways to perform pooling to condense the information; in our example we will use the method called *max-pooling*. This consists of condensing the information by just retaining the maximum value in the 2x2 region considered.

As mentioned above, the convolutional layer consists of many kernels and, therefore, we will apply *max-pooling* to each of those separately. In general, there can be many layers of pooling and convolutions.

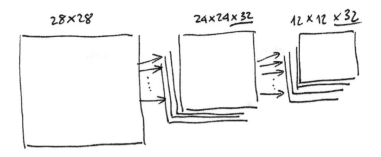

This leads that the 24x24 convolution result is transformed to a 12x12 space by the max-pooling layer that correspond to the 12x12 tiles, of which each originates from a 2x2 region. Note that, unlike in the convolutional layer, the data is tiled and not created by a sliding window.

Intuitively, we can explain *max-pooling* as finding out if a particular feature is present anywhere in the image, the exact location of the feature is not as important as the relative location with respect to others features.

Implementation of the model

In this section, I will present the example code on how to write a CNN based on the advanced example (Deep MNIST for experts) that can be found on the TensorFlow[41]website.

---

[41] TensorFlow, (2016) *Tutorials: Deep MNIST for experts*. [on line]. Availbile at: https://www.tensorflow.org/versions/master/tutorials/mnist/pros/index.html [Consulted on: 2/1/2016]

As I said in the beginning, there are many details of the parameters that require a more detailed treatment and theoretical approach than the given in this book. I hence will only give an overview of the code without going into to many details of the TensorFlow parameters.

As we have already seen, there are several parameters that we have to define for the convolution and pooling ers. We will use a stride of size 1 in each dimension (this is the step size of the sliding window) and a zero padding model. The pooling that we will apply will be a max-pooling on block of 2x2. Similar to above, I propose using the following two generic functions to be able to write a cleaner code that involves convolutions and max-pooling.

```
def conv2d(x, W):
    return tf.nn.conv2d(x, W, strides=[1, 1, 1, 1], padding='SAME')


def max_pool_2x2(x):
    return tf.nn.max_pool(x, ksize=[1, 2, 2, 1],
                strides=[1, 2, 2, 1], padding='SAME')
```

Now it is time to implement the first convolutional layer followed by a pooling layer. In our example we have 32 filters, each with a window size of 5x5. We must define a tensor to hold this weight matrix W with the shape [5, 5, 1, 32]: the first two dimensions are the size of the window, and the third is the amount of channels, which is 1 in our case. The last one defines how many features we want to use. Furthermore, we will also need to define a bias for every of 32 weight matrices. Using the previously defined functions we can write this in TensorFlow as follows:

```
W_conv1 = weight_variable([5, 5, 1, 32])
b_conv1 = bias_variable([32])
```

The ReLU (Rectified Linear unit) activation function has recently become the default activation function used in the hidden layers of *deep neural networks*. This simple function consist of returning *max(0, x)*, so it return 0 for negative values and x otherwise. In our example, we will use this activation function in the hidden layers that follow the convolution layers.

The code that we are writing will first apply the convolution to the input images *x_image*, which returns the results of the convolution of the image in a 2D tensor *W_conv1* and then it sums the bias to which finally the ReLU activation function is applied. As a final step, we apply *max-pooling* to the output:

```
h_conv1 = tf.nn.relu(conv2d(x_image, W_conv1) + b_conv1)

h_pool1 = max_pool_2x2(h_conv1)
```

When constructing a deep neural network, we can stack several layers on top of each other. To demonstrate how to do this, I will create a secondary convolutional layer with 64 filters with a 5x5 window. In this case we have to pass 32 as the number of channels that we need as that is the output size of the previous layer:

```
W_conv2 = weight_variable([5, 5, 32, 64])
b_conv2 = bias_variable([64])

h_conv2 = tf.nn.relu(conv2d(h_pool1, W_conv2) + b_conv2)
h_pool2 = max_pool_2x2(h_conv2)
```

The resulting output of the convolution has a dimension of 7x7 as we are applying the 5x5 window to a 12x12 space with a *stride* size of 1. The next step will be to add a fully connected layer to 7x7 output, which will then be fed to the final *softmax* layer like we did in the previous chapter.

We will use a layer of 1024 neurons, allowing us to to process the entire image. The tensors for the weights and biases are as follows:

```
W_fc1 = weight_variable([7 * 7 * 64, 1024])
b_fc1 = bias_variable([1024])
```

Remember that the first dimension of the *tensor* represents the 64 filters of size 7x7 from the second convolutional layer, while the second parameter is the amount of neurons in the layer and is free to be chosen by us (in our case 1024).

Now, we want to flatten the *tensor* into a vector. We saw in the previous chapter that the *softmax* needs a flattened image in the form as a vector as input. This is achieved by multiplying the weight matrix *W_fc1* with the flattend vector, adding the bias *b_fc1* after wich we apply the *ReLU* activation function.

```
h_pool2_flat = tf.reshape(h_pool2, [-1, 7*7*64])

h_fc1 = tf.nn.relu(tf.matmul(h_pool2_flat, W_fc1) + b_fc1)
```

The next step will be to reduce the amount of effective parameters in the neural network using a technique called *dropout*. This consists of removing nodes and their incoming and outgoing connections. The decision of which neurons to drop and which to keep is decided randomly. To do this in a consistent manner we will assign a probability to the neurons being dropped or not in the code.

Without going into too many details, *dropout* reduces the risk of the model of *overfitting* the data. This can happen when the hidden layers have large amount of neurons and thus can lead to very expressive models; in this case it can happen that random noise (or error) is modelled. This is known as *overfitting*, which is more likely if the model has a lot of parameters compared to the dimension of the input. The best is to avoid this situation, as overfitted models have poor predictive performance.

In our model we apply *dropout*, which consists of using the function *dropout tf.nn.dropout* before the final *softmax* layer. To do this we construct a *placeholder* to store the probability that a neuron is maintained during dropout.

```
keep_prob = tf.placeholder("float")
h_fc1_drop = tf.nn.dropout(h_fc1, keep_prob)
```

Finally, we add the *softmax* layer to our model like have been done in the previous chapter. Remember that *sofmax* returns the probability of the input belonging to each class (digits in our case) so that the total probability adds up to 1. The *softmax* layer code is as follows:

```
W_fc2 = weight_variable([1024, 10])
b_fc2 = bias_variable([10])

y_conv=tf.nn.softmax(tf.matmul(h_fc1_drop, W_fc2) + b_fc2)
```

Training and evaluation of the model

We are now ready to train the model that we have just defined by adjusting all the weights in the convolution, and fully connected layers to obtain the predictions of the images for which we have a label. If we want to know how well our model performs we must follow the example set in the previous chapter.

The following code is very similar to the one in the previous chapter, with one exception: we replace the *gradient descent optimizer* with *the ADAM optimizer,* because this algorithm implements a different optimizer that offers certain advantages according to the literature[42].

We also need to include the additional parameter *keep_prob* in the *feed_dict* argument, which controls the probability of the *dropout* layer that we discussed earlier.

```
cross_entropy = -tf.reduce_sum(y_*tf.log(y_conv))
train_step = tf.train.AdamOptimizer(1e-
4).minimize(cross_entropy)
correct_prediction = tf.equal(tf.argmax(y_conv,1), tf.argmax(y_,1))
accuracy = tf.reduce_mean(tf.cast(correct_prediction, "float"))

sess = tf.Session()

sess.run(tf.initialize_all_variables())
```

---

[42] TensorFlow, (2016) *Python API. ADAM Optimizer* [on líne]. Available at: https://www.tensorflow.org/versions/master/api_docs/python/train.html#AdamOptimizer [Accessed: 2/1/2016].

```
for i in range(20000):
  batch = mnist.train.next_batch(50)
  if i%100 == 0:
    train_accuracy = sess.run( accuracy, feed_dict={
      x:batch[0], y_: batch[1], keep_prob: 1.0})
    print("step %d, training accuracy %g"%(i, train_accuracy))
  sess.run(train_step,feed_dict={x: batch[0], y_: batch[1],
keep_prob: 0.5})

print("test accuracy %g"% sess.run(accuracy, feed_dict={
  x: mnist.test.images, y_: mnist.test.labels, keep_prob: 1.0}))
```

Like in the previous models, the entire code can be found on the *Github* page of this book, one can verify that this model achieves 99.2% accuracy.

Here is when the brief introduction to building, training and evaluating *deep neural networks* using TensorFlow comes to an end. If the reader have managed to run the provided code, he or she has noticed that the training of this network took noticeably longer time than the one in the previous chapters; you can imagine that a network with much more layers will take a lot longer to train. I suggest you to read the next chapter, where is explained how to use a GPU for training, which will vaslty decrease your training time.

The code of this chapter can be found in *CNN.py* on the *github* page of this book[43], for studying purposes the code can be found in its entirity below:

---

[43] *Github*, (2016) Source code of this book [on líne]. Availible at:
https://github.com/jorditorresBCN/TutorialTensorFlow
[Consulted on: 29/12/2015].

```
import input_data
mnist = input_data.read_data_sets('MNIST_data', one_hot=True)
import tensorflow as tf


x = tf.placeholder("float", shape=[None, 784])
y_ = tf.placeholder("float", shape=[None, 10])

x_image = tf.reshape(x, [-1,28,28,1])
print "x_image="
print x_image


def weight_variable(shape):
 initial = tf.truncated_normal(shape, stddev=0.1)
 return tf.Variable(initial)


def bias_variable(shape):
 initial = tf.constant(0.1, shape=shape)
 return tf.Variable(initial)



def conv2d(x, W):
 return tf.nn.conv2d(x, W, strides=[1, 1, 1, 1], padding='SAME')


def max_pool_2x2(x):
  return tf.nn.max_pool(x, ksize=[1, 2, 2, 1],
            strides=[1, 2, 2, 1], padding='SAME')



W_conv1 = weight_variable([5, 5, 1, 32])
b_conv1 = bias_variable([32])

h_conv1 = tf.nn.relu(conv2d(x_image, W_conv1) + b_conv1)
h_pool1 = max_pool_2x2(h_conv1)

W_conv2 = weight_variable([5, 5, 32, 64])
b_conv2 = bias_variable([64])
```

```
h_conv2 = tf.nn.relu(conv2d(h_pool1, W_conv2) + b_conv2)
h_pool2 = max_pool_2x2(h_conv2)

W_fc1 = weight_variable([7 * 7 * 64, 1024])
b_fc1 = bias_variable([1024])

h_pool2_flat = tf.reshape(h_pool2, [-1, 7*7*64])
h_fc1 = tf.nn.relu(tf.matmul(h_pool2_flat, W_fc1) + b_fc1)

keep_prob = tf.placeholder("float")
h_fc1_drop = tf.nn.dropout(h_fc1, keep_prob)

W_fc2 = weight_variable([1024, 10])
b_fc2 = bias_variable([10])

y_conv=tf.nn.softmax(tf.matmul(h_fc1_drop, W_fc2) + b_fc2)

cross_entropy = -tf.reduce_sum(y_*tf.log(y_conv))
train_step = tf.train.AdamOptimizer(1e-
4).minimize(cross_entropy)
correct_prediction = tf.equal(tf.argmax(y_conv,1), tf.argmax(y_,1))
accuracy = tf.reduce_mean(tf.cast(correct_prediction, "float"))


sess = tf.Session()

sess.run(tf.initialize_all_variables())

for i in range(200):
  batch = mnist.train.next_batch(50)
  if i%10 == 0:
    train_accuracy = sess.run( accuracy, feed_dict={
      x:batch[0], y_: batch[1], keep_prob: 1.0})
    print("step %d, training accuracy %g"%(i, train_accuracy))
  sess.run(train_step,feed_dict={x: batch[0], y_: batch[1],
keep_prob: 0.5})

print("test accuracy %g"% sess.run(accuracy, feed_dict={
  x: mnist.test.images, y_: mnist.test.labels, keep_prob: 1.0}))
```

6. Parallelism

The first TensorFlow package, appearing in November 2015, was ready to run on servers with available GPUs and executing the training operation simultaneously in them. In February 2016, an update added the capability to distribute and parallelize the processing.

In this short chapter I'll introduce how to use the GPUs. For those readers wanting to understand a little bit more of how these devices work, some references will be given in the last section, but. Given the introductory focus of this book, I'll not enter in detail for the distributed versión, but for those readers interested some references will be given in the last section.

Execution environment with GPUs

The TensorFlow package supporting GPUs requires the *CudaToolkit 7.0* and *CUDNN 6.5 V2*. For installing the environment, we suggest to visit the *cuda installation*[44] website, for not going deep in details, also the information is up-to-date.

---

[44] TensorFlow, (2016) GPU-related issues. [online]. Available at: https://www.tensorflow.org/versions/master/get_started/os_setup .html#gpu-related-issues [Accessed: 16/12/2015].

The way to reference those devices in TensorFlow is the following one:

- "/cpu:0": To reference the server's CPU.
- "/gpu:0": The server's GPU, if only one is available.
- "/gpu:1": The second server's GPU, and so on.

To know in which devices our operations and tensors are assigned we need to create a sesion with the option *log_device_placement* as *True*. Let's see it in the following example:

```
import tensorflow as tf
a = tf.constant([1.0, 2.0, 3.0, 4.0, 5.0, 6.0], shape=[2, 3], name='a')
b = tf.constant([1.0, 2.0, 3.0, 4.0, 5.0, 6.0], shape=[3, 2], name='b')
c = tf.matmul(a, b)

sess =
tf.Session(config=tf.ConfigProto(log_device_placement=True))
printsess.run(c)
```

When the reader tests this code in their computer, a similar output should appear:

```
. . .
Device mapping:
/job:localhost/replica:0/task:0/gpu:0 -> device: 0, name: Tesla
K40c, pci bus id: 0000:08:00.0
. . .
b: /job:localhost/replica:0/task:0/gpu:0
a: /job:localhost/replica:0/task:0/gpu:0
MatMul: /job:localhost/replica:0/task:0/gpu:0
...
 [[ 22.28.]
 [ 49.64.]]
...
```

Also, with the result of the operation, it informs to us where is executed each part.

If we want a specific operation to be executed in a specific device, instead of letting the system select automatically a device we can use the variable *tf.device* to create a device context, so all the operations in that context will have the same device assigned.

If we have more that a GPU in the system, the GPU with the lower identifier will be selected by default. In case that we want to execute operations in a different GPU, we have to specify this explicitly. For example, if we want the previous code to be executed in GPU #2 we can use *tf.device('/gpu:2')* as shown here:

```
import tensorflow as tf

with tf.device('/gpu:2'):
a = tf.constant([1.0, 2.0, 3.0, 4.0, 5.0, 6.0], shape=[2, 3], name='a')
b = tf.constant([1.0, 2.0, 3.0, 4.0, 5.0, 6.0], shape=[3, 2], name='b')
c = tf.matmul(a, b)
sess =
tf.Session(config=tf.ConfigProto(log_device_placement=True))
printsess.run(c)
```

Parallelism with several GPUs

In case that we have more tan one GPU, usually we'd like to use all together to solve the same problem parallelaly. For this, we can build our model to distribute the work among several GPUs. We see it in the next example:

```
import tensorflow as tf

c = []
for d in ['/gpu:2', '/gpu:3']:
with tf.device(d):
a = tf.constant([1.0, 2.0, 3.0, 4.0, 5.0, 6.0], shape=[2, 3])
b = tf.constant([1.0, 2.0, 3.0, 4.0, 5.0, 6.0], shape=[3, 2])
c.append(tf.matmul(a, b))
with tf.device('/cpu:0'):
sum = tf.add_n(c)

# Creates a session with log_device_placement set to True.
sess =
tf.Session(config=tf.ConfigProto(log_device_placement=True))
print sess.run(sum)
```

As we see, the code is equivalent to the previous one but now we have 2 GPUs indicated with *tf.device* performing a multiplication (both GPUs do the same here, to simplify the example code), and later the CPU performs the addition. Given that we set *log_device_placement* as *true*, we can see in the output how the operations are distributed through our devices[45].

```
...
Device mapping:
/job:localhost/replica:0/task:0/gpu:0 -> device: 0, name: Tesla K40c
/job:localhost/replica:0/task:0/gpu:1 -> device: 1, name: Tesla K40c
```

---

[45]   This output is result of using a server with 4 Tesla K40 GPUs from the Barcelona Supercomputing Center (BSC-CNS).

```
/job:localhost/replica:0/task:0/gpu:2 -> device: 2, name: Tesla K40c
/job:localhost/replica:0/task:0/gpu:3 -> device: 3, name: Tesla K40c
...


...
Const_3: /job:localhost/replica:0/task:0/gpu:3
I tensorflow/core/common_runtime/simple_placer.cc:289] Const_3:
/job:localhost/replica:0/task:0/gpu:3
Const_2: /job:localhost/replica:0/task:0/gpu:3
I tensorflow/core/common_runtime/simple_placer.cc:289] Const_2:
/job:localhost/replica:0/task:0/gpu:3
MatMul_1: /job:localhost/replica:0/task:0/gpu:3
I tensorflow/core/common_runtime/simple_placer.cc:289] MatMul_1:
/job:localhost/replica:0/task:0/gpu:3
Const_1: /job:localhost/replica:0/task:0/gpu:2
I tensorflow/core/common_runtime/simple_placer.cc:289] Const_1:
/job:localhost/replica:0/task:0/gpu:2
Const: /job:localhost/replica:0/task:0/gpu:2
I tensorflow/core/common_runtime/simple_placer.cc:289] Const:
/job:localhost/replica:0/task:0/gpu:2
MatMul: /job:localhost/replica:0/task:0/gpu:2
I tensorflow/core/common_runtime/simple_placer.cc:289] MatMul:
/job:localhost/replica:0/task:0/gpu:2
AddN: /job:localhost/replica:0/task:0/cpu:0
I tensorflow/core/common_runtime/simple_placer.cc:289] AddN:
/job:localhost/replica:0/task:0/cpu:0
[[44.56.]
 [98.128.]]
...
```

Code example with GPUs

To conclude this brief chapter, we present a snippet of code inspired on the one shared by DamienAymeric in *Github*[46],

---

[46] *Github* (2016) AymericDamien. [online]. Available at:
https://github.com/aymericdamien/TensorFlow-Examples

computing A^n+B^n for $n=10$ comparing the execution time with 1 GPU against 2 GPUs, using the *datetime* Python package.

First of all, we import the required libraries:

```
import numpy as np
import tensorflow as tf
import datetime
```

We create two matrix with random values, using the *numpy* package:

```
A = np.random.rand(1e4, 1e4).astype('float32')
B = np.random.rand(1e4, 1e4).astype('float32')

n = 10
```

Then, we create the two structures to store the results:

```
c1 = []
c2 = []
```

Next, we define the *matpow()* function as follows:

```
defmatpow(M, n):
    if n < 1: #Abstract cases where n < 1
        return M
    else:
```

[Accessed: 9/1/2015].

```
return tf.matmul(M, matpow(M, n-1))
```

As we've seen, to execute the code in a single GPU, we have to specify this as follows:

```
with tf.device('/gpu:0'):
    a = tf.constant(A)
    b = tf.constant(B)
    c1.append(matpow(a, n))
    c1.append(matpow(b, n))

with tf.device('/cpu:0'):
  sum = tf.add_n(c1)

t1_1 = datetime.datetime.now()

with
tf.Session(config=tf.ConfigProto(log_device_placement=True)) as
sess:
sess.run(sum)
t2_1 = datetime.datetime.now()
```

And for the case with 2 GPUs, the code is as follows:

```
with tf.device('/gpu:0'):
    #compute A^n and store result in c2
    a = tf.constant(A)
    c2.append(matpow(a, n))

#GPU:1
with tf.device('/gpu:1'):
    #compute B^n and store result in c2
    b = tf.constant(B)
    c2.append(matpow(b, n))

with tf.device('/cpu:0'):
    sum = tf.add_n(c2) #Addition of all elements in c2, i.e. A^n +
B^n
```

```
t1_2 = datetime.datetime.now()
with
tf.Session(config=tf.ConfigProto(log_device_placement=True)) as
sess:
    # Runs the op.
sess.run(sum)
t2_2 = datetime.datetime.now()
```

Finally we print the results for the registered computation time:

```
print "Single GPU computation time: " + str(t2_1-t1_1)
print "Multi GPU computation time: " + str(t2_2-t1_2)
```

Distributed version of TensorFlow

As I previously said at the beginning of this chapter, on February 2016 Google released the distributed version of TensorFlow, which is supported by gRPC, a high performance open source RPC framework for inter-process communication (the same protocol used by TensorFlow Serving).

For its usage, the binaries must be built because the package only provides the sources at this time. Given the introductory scope of this book, I will not explain it in the distributed version, but if the reader wants to learn about it, I recommend to start with the oficial site for this distributed version of TensorFlow[47].

---

[47]   Distributed TensorFlow, (2016) [online]. Available at: https://github.com/tensorflow/tensorflow/tree/master/tensorflow/core/distributed_runtime [Accessed: 16/12/2015].

As in previous chapters, the code used in this one, can be found in the book's *Github*[48]. I hope that this chapter has been illustrative enough to understand how to speedup the code using GPUs.

---

[48] *Github*, (2016) Source code of this book [on líne]. Availible at: https://github.com/jorditorresBCN/TutorialTensorFlow [Consulted on: 29/12/2015].

Closing

Exploration is the engine that drives innovation. Innovation drives economic growth. So let's all go exploring.
Edith Widder

Here I have presented an introductory guide explaining how to use TensorFlow, providing a warm-up with this technology that will undoubtedly have a leading role in the looming technological scenario. There are, indeed, other alternatives to TensorFlow, and each one suit the best a particular problem; I want to invite the reader to explore beyond the TensorFlow package.

There is lots of diversity in these packages. Some are more specialized, others less. Some are more difficult to install than others. Some of them are very well documented while others , despite working well, are more difficult to find detailed information about how to use them.

An important thing: on the following day that TensorFlow was release by Google, I read in a *tweet*[49] that during the period 2010-2014 a new *Deep learning* package was released every 47 days, and in 2015 releases were published every 22 days. It is mpressive, isn't it? As I advanced in the first chapter of the book, as a starting point for the reader, an extensive list can be found at *Awesome Deep Learning*[50].

Without any doubt, the landscape of *Deep Learning* was impacted in November 2015 with the release of Google's TensorFlow, being now the most popular open source machine learning library on *Github* by a wide margin[51].

Remember that the second most-starred *Machine Learning* project of *Github* is Scikit-learn[52], the *de facto* official Python general *Machine Learning* framework. For these users, TensorFlow can be used through Scikit Flow (skflow)[53], a simplified interface for TensorFlow coming out from Google.

---

[49] Twitter (11/11/2015). Kyle McDonald: *2010-2014: new deep learning toolkit is released every 47 days. 2015: every 22 days.* [Online]. Available at: https://twitter.com/kcimc/status/664217437840257024 [Accessed: 9/01/2016].

[50] *GitHub,* (2016) *Awesome Deep Learning.* [Online]. Available at: https://github.com/ChristosChristofidis/awesome-deep-learning [Accessed: 9/01/2016].

[51] Explore GitHub, Machine learning: [Online]. Available at: https://github.com/showcases/machine-learning [Accessed on: 2/01/2016]

[52] Scikit-Learn GitHub: [Online]. Available at: https://github.com/scikit-learn/scikit-learn [Accessed: 2/3/2016]

[53] Tensorflow/skflow GitHub: [Online]. Available at: https://github.com/tensorflow/skflow [Accessed: 2/1/2016]

Practically, Scikit Flow is a high level wrapper for the TensorFlow library, which allows the training and fitting of neural networks using the familiar approach of Scikit-Learn. This library covers a variety of needs from linear models to *Deep Learning* applications.

In my humble opinion, after the release of the distributed version of TensorFlow, TensorFlow Serving and Scikit Flow, TensorFlow will become *de facto* a mainstream *Deep Learning* library.

Deep learning has dramatically improved the state-of-the-art in speech recognition, visual object recognition, object detection and many other domains. What will be its future? According to an excellent review from Yann LeCun, Yoshua Bengio and Geoffrey Hilton in Nature journal, the answer is the *Unsupervised Learning* [54]. They expect *Unsupervised Learning* to become far more important in longer term than supervised learning. As they mention, human and animal learning is largely unsupervised: we discover the structure of the world by observing it, not by being told the name of every object.

They have a lot of espectations of the future progress of the systems that combine CNN with recurrent neural network (RNN) that use reinforcement learning. RNNs process an input that sequence one element at time, maintaining in their hidden units information about the history of all the past elements of the sequence. For an introduction to RNN im-

---

[54] Yann LeCun, Yoshua Bengio and Geoffrey Hinton (2015). "Deep Learning". Nature 521: 436–444 doi:10.1038/nature14539. Available at: http://www.nature.com/nature/journal/v521/n7553/full/nature14539.html.

plementation in TensorFlow the reader can review the *Recurrent Neural Networks* [55] section in TensorFlow Tutorial.

Besides, there are many challenges ahead in *Deep Learning*; the time to training them are driving the need for a new class of supercomputer systems. A lot of research is still necessary in order to integrate the best analytics knowledge with new Big Data technologies and the awesome power of emerging computational systems in order to interpret massive amounts of heterogeneous data at an unprecedented rate.

Scientific progress is typically the result of an interdisciplinary, long and sustained effort by a large community rather than a breakthrough, and deep learning, and machine learning in general, is not an exception. We are entering into an extremely exciting period for interdisciplinary research, where ecosystems like the ones found in Barcelona as UPC and BSC-CNS, with deep knowledge in *High Performance Computing* and Big Data Technologies, will play a big role in this new scenario.

---

[55] TensorFlow, (2016) Tutorial – Recurrent Neural Networks [Online]. Available at: https://www.tensorflow.org/versions/ r0.7/tutorials/recurrent/index.html [Accessed: 16/03/2016].

Acknowledgments

First of all I want to thanks everybody that helped me to write the Spanish version of this book. All of them are mentioned in the Acknowledgments of the Spanish book[56].

For this English version, I want to give special thanks to my colleagues that helped me with the translation of this book: Joan Capdevila, Mauro Gómez, Josep Ll. Berral, Katy Wallace and Christopher Bonnett. Without their great support I couldn't have finished this English version before Easter.

Finally, indicate that this work is partially supported by the grant SEV-2015-0493 of Severo Ochoa Program awarded by the Spanish Government, the TIN2015-65316-P project, with funding from the Spanish Ministry of Economy and Competitivity, the European Union FEDER funds, and the SGR 2014-SGR-1051.

---

[56] Hello World en TensorFlow [Online]. Available at: http://www.jorditorres.org/libro-hello-world-en-tensorflow/ [Accessed: 16/03/2016].

About the Author

Jordi Torres is a professor at UPC Barcelona Tech and a research manager and senior advisor at Barcelona Supercomputing Center with a wide range of research and teaching activities for over 25 years. With a great background as a Computer Engineer, his explorer and entrepreneurial spirit have led him to be a Big-Data engineer able to engage with Data Scientists. Actually, his research focus is gradually moving from supercomputing architectures and runtimes to execution middleware's for big data workloads, and more recently to platforms for Machine Learning on massive data. Right now, he also has a consultative and strategy role with a visionary task related to next generation technology and its impact. He is both a creative thinker and influential collaborator, and for that he has worn many hats throughout his long career. He acts as an expert for various organizations and companies and mentors entrepreneurs. He is also a writer, gives conferences and collaborates with Spanish mass media. He is passionate about art and visual design. More information about him can be found on his web page *http://www.JordiTorres.Barcelona*

About BSC

Barcelona Supercomputing Center (BSC) is the leading supercomputing center in Spain. It specialises in High Performance Computing (HPC) and its mission is two-fold: to provide infrastructure and supercomputing services to
European scientists, and to generate knowledge and technology to transfer to business and society. BSC is a Severo
Ochoa Center of Excellence and a first level hosting member
of the European research infrastructure PRACE (Partnership
for Advanced Computing in Europe). BSC also manages the
Spanish Supercomputing Network (RES). More information
on page *www.bsc.es*

About UPC

The Universitat Politècnica de Catalunya · BarcelonaTech
(UPC) is a public institution dedicated to higher education
and research, specialised in the fields of engineering, architecture and science. The activity that goes on at UPC cam-
puses and schools has made the University a benchmark institution. The University harnesses the potential of basic and
applied research, and transfers technology and knowledge to
society. As a leading member of international networks of
excellence, the UPC has a privileged relationship with global
scientific and educational organisations. More information
on page *www.upc.edu*

About Grup d'Estudi de Machine Learning de Barcelona

The Grup d'Estudis de Machine Learning de Barcelona (GEMLeB) is an organization devoted to understand, discover and promote the use of Machine Learning in Barcelona, through an informal environment. As many other machine learning meetups, regular meetings are organized with a two-fold objective: learn about machine learning (from experiences, applications to algorithms, models and theory) and meet people with similar interest to build a wide and supporting community. If you ever come to Barcelona and want to meet the community, please do not hesitate to contact us! More information on page *www.meetup.com/Grup-destudi-de-machine-learning-de-Barcelona*

Printed in Great Britain
by Amazon